A NOVEL

MAJOR GIGOLO

TOM E. QUINN

ISBN
978-1-958690-32-1 (Paperback)
978-1-958690-33-8 (eBook)
978-1-958690-31-4 (Hardcover)

Dedication

This book is dedicated to all the military, government and contracted employees of the Department of Defense. Everyday they toil in the trenches of both real and an unrealized war. For all the information warriors that understand middle management. Management that has usurped, copied or downright plagiarized some damn good work, and never even said "Thanks, take tomorrow off".
It is for these folks that I hope to give them few pages of entertainment.

To the officers, soldiers, sailors, and troopers in the US Pentagon; I especially want to say, "THANK YOU" for your everyday work to keep your country free and the kids fighting forward alive. I absolutely respect and admire your dedication to duty and your humble service.

To my friends that said I needed to put more sex in my books. Sometimes, you really do get what you ask for!

TABLE OF CONTENTS

ACKNOWLEDGEMENTS

I would like to thank all the soldiers, sailors, airmen and Marines that fight to keep this country free in the face of countless, faceless and merciless enemies of freedom.

Thanks to all the people in the Pentagon that still work for defense of the nation and don't let politics get in the way of the mission.

To my family for sticking with me through the years. The separations were tough, but you know I Love You, and that's all that matters.

PREFACE

This novel is about change. It is symbolic of the perpetual change of the U. S. military. What was once the greatest defense force for the stability on the globe has changed into an offensive force that has generated instability and lost the moral high ground in the process. America's leaders have directed that change from defensive to offensive posture to prosecute the War on Terror. A change I wholeheartedly concur with, yet understand the concern of my countrymen.

I agree with the War on Terror, but I am not blind to the other challenges our military face around the globe. It is my point of view that this change was not good for the fighting warriors of the American military. Certainly, America's soldiers will fight. They love to fight. That's what our adversaries fear most about us. Our enemies cannot meet us on a battlefield so they use tactics that they feel will defeat us.

However those that hate us do not understand us any more than we understand them. Freedom can never be defeated. They will never understand this. But our enemies are a patient lot. We, my friends, are not. Thus, the need for change.

I am a proponent for War. War, with all it's horror and tragedy is a cleansing. It is God's way of giving us, mankind, more rope with which to hang ourselves. War does, however brutal it may be, solve problems.

Our servicemen are not being allowed to solve the problems we face. As American politicians dicker with thoughts of diplomacy as an alternative to War, American warriors perform admirably in a sort of military masturbation called 'Peacekeeping' all over the planet. Neither allowed to pursue victory, nor accepting defeat. Caught in a real life purgatory, while politicians pontificate.

My hero, Major Ron Benson represents my perspective of the American fighting man in the 21st Century. A warrior longing to be turned loose to fight in a War for his country, yet unable to. What other option would a warrior want? Perhaps love?

The hero in the book was not easy for me to create. It was a difficult decision to make a military officer a gigolo. It was not difficult to put the gigolo in Washington DC. Our nation's capital is the epicenter of what's wrong with our country today. Military members consistently are ranked as trustworthy by the American public. Unlike the elected officials that oversee the military. I have only marginal respect for politicians and minimal respect for our congressional 'leaders'.

Major Ron Benson is a creation of my personal frustration. Frustration with the unacceptable, political manipulation of the finest men and women our nation has. A military manipulated by both sides of the political spectrum. The people elected politicians to lead the country forward. Not manage our tax money or quibble over power, or plunder our national wealth, resources and "the best our country has to offer" without positive communication and unity of effort.

If the government won't turn them loose to fight, I suggest they come home to fornicate. After all, if they are our "best" let's bring them home to make families. Quite possibly some of them will get into politics and not be so carefree about sending our military off to War without fully understanding the consequences and secondary effects.

Please accept "Major Gigolo" for what it is; A book of symbolic change and not the diatribe of a grumpy, disgruntled, horny old man.

Don't read too much into the symbolism. You might miss some of the fun if you look too deep.

I hope the warriors forgive me for any appearance of cheapening their honorable service to God and country. In no way, shape or form did I mean to cheapen their service.

THANK YOU for keeping us all safe.

Much respect and trust for the warriors and those that love them,

Tom E. Quinn

CHAPTER 1

16 SEPTEMBER
2D FLOOR, E RING, ROOM 800
PENTAGON, WASHINGTON DC.

Major Ron Benson looked around his cubicle at the awards and plaques and it dawned on him. His life couldn't get any worse. He truly was the master of the six by eight-foot domain that had become his life. He was the King of the Pentagon Butt-Boys.

Ronald Benson wondered to himself how much longer he would suffer inside the walls of the Pentagon. Surely when he got promoted to Lieutenant Colonel he could leave. Some General or maybe one of the semi-intelligent congressmen across the river would see his value and put him back in charge of troops.

But Ron knew better. His current life was like that of a medieval serf. Physically he worked in a six foot by eight-foot space for General Officers and Department of the Army civilians, but virtually, he planted seeds in their minds in the hopes they would grow. Take root and eventually be harvested into actions that would serve the common good of his fellow soldiers. The ultimate goal was to be recognized as the farmer that planted the seed. Then he could move to a new, larger plot of physical space. Perhaps ten foot by ten foot. With a window. A window that looked out over the Potomac. Where he could see what kind of day it was outside. Such are the dreams of Pentagon warriors. The only window Ron looked at was a Microsoft one.

For some reason, he smiled. Thinking maybe the god of information would be kind to him. Maybe there wouldn't be some tasking handed down to plant a new crop. He had only been gone for an hour. The god of information would surely be kind to him. But it was not to be. There were thirty-seven new emails. A quick glance led him to believe that only two or three were jokes. No less than five were from his boss, Colonel Henderson. He did see some front good friends that were overseas in Iraq and Afghanistan. He quickly read those and got his personal information update from the front lines. It was surprising how much the guys forward were positive about what they were doing. They were making a difference. Ron looked at his computer and frowned. *I won't be making a difference until I get out of the Pentagon.*

He sat back on his virtual farm and cursed the demon. "Fuck you, Bill Gates. Did you really know what you were doing?"

His question was almost overheard as his cubicle-mate entered behind him. He was beaming. His co-serfs words also implied the tragedy of the life they led, but were not nearly as philosophical as Ron's thoughts.

"Being a Junior Officer and briefing at the Pentagon, is like an ice water enema!" declared Major Tony Parkman. "You know what's gonna happen and you know it's gonna suck. But when it's over, damn you feel good!" The Air Force Major seemed impressed with his gross analogy. He held up a coin he was given just minutes before and looked at it as if it were pure gold. Coins were the 21st Century answer to medals. They were presented to troops, or in this case, a staff officer for doing something above and beyond the ordinary. It was a way for a senior officer to recognize someone instantly versus the normal six-month wait it took to give out a medal. Major Parkman had been recognized for providing the US Air Force Chief of Staff an outstanding 32 slide presentation on why Strategic Communications was a "Center of Gravity" for the Global Combatant Commanders. Another seed was planted. The General's coin was a way to tell the serf his crop might grow. Major Ron Benson would not be getting a coin, or a Certificate or anything else any time soon.

Ron leaned back in his chair and shook his head as Tony's comment sank in. "I think the whole ice water enema thing is a bit more than I need to know. That is a joke right? I mean you never really had one?"

Tony dismissed the question and jabbed his friend's arm. "Seriously, I owe this thing to you, Ron." Tony held the coin up and showed it to his peer. "That brief was a piece of art, Ron. A Pentagon Picasso."

Ron already knew the briefing would go well. He had written it. To complete the crop, it needed to be planted by a slick talking serf. That was where Tony came in.

Tony Parkman, not quite 35, still looked like a male model. Standing six foot two, with blond hair long enough to push the standards of military regulations, a body that used to be chiseled by weightlifting two hours a day was starting to show the initial signs of the staff officer waist with love handles on the flanks. He was an F15 pilot that was performing his required staff work as the Air Force groomed him for command. Tony Parkman was an anomaly as a serf on the virtual farm. He was only marking time on his virtual plot. He was destined to leave the farm and separate himself from the other serfs. He was also, the epitome of what the Pentagon had become. All show, no substance.

At the opposite end of the spectrum was Ron Benson, the ultimate serf. He was no model. On a good day, he could stretch to six feet tall. He was starting to get a little roll around his stomach and he felt his pants were getting switched at the dry cleaners with someone much smaller. His brown hair was beginning to turn gray, but everyday he was thankful he still had it all. Plenty of other majors in the building were already going bald. His blue eyes could still shine, but they seemed to be bloodshot most of the time. Some wrinkles were starting to form on his face. His was a face that appeared too young to look so old.

Ron was undoubtedly the senior ranking Major in the entire Pentagon. He had been passed over for promotion to Lieutenant Colonel three times and knew he only had one more shot at making the next grade. The promotion would be his ticket to a new job, probably commanding a battalion. He was due to find out the results of the latest promotion board any day. At 40 years old and with over 18 years of service, Ron just wanted to hold on to rank to make his 20 years of service and retire. He had almost given up on a promotion. He had actually hoped the feudal lords would be kind to him this year. Reality was always his strong point so his goal was much lower. He only wanted to serve his country as best he could. Certainly leading troops was the pinnacle for any officer, but at this point, Ron would take any job that got him close to troops. Every self-respecting officer would give his left nut to lead troops. For Ron, any hope of being with troops and fighting the war was fading with each passover.

Ron smiled at Tony. "At least they know your name". Major Ron Benson wasn't sure that anyone had known his name anymore. At that particular moment, that wasn't a bad thing. Notoriety at the Pentagon promised only an increased workload and unwanted taskers. Neither of which Ron Benson wanted any part of.

For either Major, the Coin was nice and well deserved, but it would have gladly been traded for a three-day pass. It was no secret that working at the Pentagon required higher than average dedication to duty. Even in the Public Affairs Division, the majority of officers seemed to work ten-hour days at a minimum.

"This coin and fifty cents will get me a cup of coffee at the cafeteria." Tony shook his head and placed the Coin down on his desk. "Speaking of which, I'm heading there for a cup o' Joe. You want one?"

"No, I'm gonna finish up this next stinking presentation I've got to get done. Colonel Henderson will have my ass if I don't."

Tony nodded, "Suit your self. See you in about thirty minutes." It took that long to get to the cafeteria, get through the line, and talk with five or six people from around the building to find out what was really happening.

Ron wasn't sure, but the cafeteria might be the place where the real work was coordinated. Maybe that was how the Pentagon really got information out to subordinates. The worker bees would share rumors that the leadership wanted to be worked on, while they kept the true important things in the Secretary of Defense's office. It had occurred to Ron a couple years earlier that he was a step below "middle management". He was a staff officer in the largest office building in the free world. A place where men and women worked daily with all their heart and soul to keep America free, yet rarely saw any results for their dedication. He had worked for 18 years to achieve the mediocre life that only a minute percentage of officers would ever have the opportunity to endure.

The few subordinates Ron had in the office worshipped him. To his peers, he was always counted on to have answers. Now he was the author of the latest offering to the 'E-gods'. His 32-slide power point briefing was floating around the puzzle palace, perhaps one day to take root.

The Generals usually chose to ignore staff recommendations because that would require a decision. Decisions would lead to actions and the

Pentagon was no place for that. Only the Secretary seemed to be allowed to make decisions, such as they were.

The boss of Ron's division was Colonel Harold Henderson. His call sign around the office was, "H2 No". Henderson was the typical Pentagon Colonel. He was actually quite capable of being promoted as he had already mastered indecisiveness. If suggestions did not meet with what the Generals wanted, Henderson would never risk his rank or retirement to support a subordinate. Such had become the 21St Century Pentagon. Home of the world's greatest military minds, restricted not by a lack of creativity or vision, but by civilian politicians with a corporate mind set that over classified secrets that weren't so secret and ran their daily lives with a ridiculous fear of making decisions.

Any suggestions brought up by a junior officer were bound to be turned down and go no further than the colonels' inbox. H2 No was the epitome of the 'Peter Principle'. He had truly reached the maximum possible level of his ability as a Lieutenant Colonel and should have gone no further. Without the vision required to achieve a generals' star, Henderson had done the best he could becoming a full bird Colonel in the United States Air Force. For Henderson, it was at least one pay grade higher than his ability.

Henderson constantly made Ron Benson prepare briefings in power point. As was Henderson's annoying habit, upon reviewing the slides, He would send the paper version back to Ron with red ink corrections. Rather than actually make the corrections on the slides while viewing them on the computer, which would save countless time, money and effort, Henderson used this method to harass his subordinates. It was his little way to make sure the serf knew the master. The worst part was that 95% of the changes were "little dog" changes. The type where he would make suggestions that were nothing more than equivalent terms. Such as "little dog" would become "puppy" or "vision" would become "perspective". The other piece that pissed Ron off was the fact that Henderson was responsible for killing at least two trees a week in the copy machine.

The effort for Ron that evening was correcting a 68 slide show that showed how the Department of Defense was going to present the F22 Strike Fighter as the primary weapon for the next twenty years. Ron Benson had many issues with the presentation, none of which agreed with the premise.

First and foremost, the Air Force was going to do and say anything to make sure the 21St Century manned fighter would be purchased. At nearly $200 million per copy, Ron knew a couple Army Divisions could be fielded for less. Secondly, Colonel Henderson would only say no to subordinates so he was going to agree with the Air Forces perspective no matter what. If an Air Force General Officer said Colonel Henderson needed to shit a jet, he would do it. Or more than likely, make some poor Major shit a jet, and then claim the jet was actually the jet he had just shit.

Ron read the red writing on the first slide:

"You make this sound too negative. You need to make this presentation sing to the Four Star that the F22 is something we can't do without!!!"

Ron shook his head in disgust. It was hard to make a slide presentation "sing" when you didn't agree with the song. As an Army Officer, he could see absolutely no reason to spend that kind of money on any one piece of equipment. If the taxpayers ever figured out what was going on with the Defense money they might actually revolt.

Ron Benson knew that wouldn't happen. He was a good enough officer to understand the mission. For years soldiers had been getting missions, tasks really, exactly like this. Taking a shit sandwich and making a meal out of it. As much as he disagreed with it, he executed it to the best of his ability.

Ninety minutes later, the task was complete. Tony had left over an hour ago. Ron was the last person in the office. He looked at his watch and saw it was after eight. Ron exhaled heavily and realized it was time to head home to his empty apartment. Sometimes he would change into civilian clothes and go out for a beer or a movie. He was beginning to become envious of the contractors that worked by the clock and not the task. Perhaps more free time would make him feel better about his life. Maybe he needed some kind of hobby.

He walked to the door and turned to see the empty office. Twenty empty cubicles for thirty people. The last one there was him. No supervisor, no co-workers, not even a janitor. He turned off the light and silently closed the door. It was a lonely ride home.

CHAPTER 2

20 SEPTEMBER
2E800
THE PENTAGON

It was the end of the day and most of Ron's co-workers were already headed home. He received an email saying, "Please come to my office at 5 o'clock." It was from Henderson. He left no indication of what the visit would be about, but Ron had learned one thing. Any visit to Henderson's office was not good. But one at the end of the day was certain to suck.

He showed up at exactly five. Henderson directed him to take a seat. Ron sat down, but no matter how hard he tried, he couldn't get comfortable. The first thing he noticed was that Colonel Henderson was ready to leave. His desk was already cleaned up and he had his coat on as he sat behind his desk. "I'm not going to beat around the bush on this, Ron. You've been passed over for promotion. I'm sorry."

Ron Benson shook his head. He was not expecting that kind of news. Nearly every other Major he knew that had been passed over three times had made it on their fourth look. Or they had already retired. Of course he had already been passed over three times before so the news was not unfamiliar. It was however, unexpected in the current setting. He was stunned. For an instant, he thought he could feel tears starting to form. This was his last chance to get picked up for Lieutenant Colonel. It was the one chance that the military powers that be may have seen the error of their ways and actually promoted someone that had been hosed by the system.

But he fought off the tears. Ron had been in the military long enough to know sometimes, a serf remains a serf. He bit his lip blink his eyes to gather himself. He would not let H2No see his pain.

He said quietly, "Is there any reason why?" He expected to have had some sort of indication this news was coming to soften the blow. He wouldn't get that from his supervisor.

Henderson feinted he was concerned. "They don't tell me the why on these sorts of things. I imagine it was the back to back Center Mass OER's you got." A Center Mass Officer Efficiency Report meant that the supervising official had considered Ron to be average. One report like that is generally not conducive to promotion. In the US Army of the 21st Century, two meant you needed to find a new line of work. Ron Benson already assumed because of those two pieces of paper, any and all work he did at the Pentagon was irrelevant for a promotion.

"If there is anything I can do for you, you know I'll do it. We put in a great report together for the last board." Ron nodded. Ron knew how good it was because he wrote it. Ron understood Henderson had tried to take care of him by signing the document and pushing it up to his boss, Major General Patrick McCoy. It was too little too late in the eyes of the Army promotion board. A Colonel, from the US Air Force was trying to save some mediocre Army Major's career. Henderson for his part was concerned about Ron personally, but didn't know the right buttons to push on the Army side to get Ron promoted. It was a nice try, but the Air Force knows about jets, not Army personnel matters.

Ron shook his head. "No sir, I don't think there is anything else."

That was what Henderson wanted to hear. His grief and pain for his inability to get his subordinate promoted was officially waived off by Ron. Henderson was late and the traffic on I-95 wasn't getting any lighter. He got up from his desk. "Again, I'm sorry, Ron." He pulled a folder from his out box. "I also need you to hit these slides one more time. I need to have them first thing tomorrow."

Ron shook his head again and smiled. Unbelievable! What balls? The son of a bitch just got through telling him in essence his career was done, and now was asking him to stay late to do work. He wanted desperately to shove the entire folder so far up Henderson's ass that he would get paper cuts on his tonsils.

That's not what soldiers do. And above it all, Ron was a soldier. He nodded his head and said, "Sure thing, sir. I'll have 'em on your desk tonight before I leave." He had to bite his tongue to keep from adding, "shithead".

Henderson handed him the folder and patted him on the shoulder. "See you tomorrow, Ron."

Ron took the folder and headed to his desk. He looked around the office at the few co-workers left there. The beautiful woman in uniform outside Henderson's office looked up from her desk. He could tell by the expression on Lieutenant Colonel Mary Chambers face that she knew. She too was an air Force officer, but her destiny was for greater things. It didn't hurt that she was single and sleeping with a two star general. Ron wondered for just a moment if he could convince her to provide him a sympathetic romp. Maybe that would pick him up. But, the officer that was still inside him knew Mary Chambers and his pent up sexual desire would form a lethal mix. Ron decided he didn't need her sympathy and walked by without a word.

Tony Parkman was at his desk next to Ron's. "What was that all about?" It was an innocent enough question and indicated Tony had no idea what H2 No wanted.

"I didn't get picked up for O-5." Ron tossed the folder on his desk.

"You gotta be shittin' me?" exclaimed Tony. He actually seemed more upset than Ron.

"Henderson had to be the first to tell me. First line supervisor and all that crap."

"Bet he enjoyed that."

Ron shook his head. "Not really. He had to get it over though so he could get on the road. Wouldn't want to get stuck in the five-thirty traffic."

Tony let the comment slide. "I'm sorry, man."

Ron took a deep breath. "I can handle it."

"Come on. Let's get outta here."

"No can do, Tony. I gotta fix that briefing." He pointed at the folder.

Tony was perplexed. "He told you, you were passed over, then gave you a briefing to fix?"

Ron laughed. "Yup." He pulled out his chair sat down heavily. "I don't know whether that's audacity or courage?"

"Whatever it is, it's fucked up!"

"Maybe so, but it'll take my mind off being passed over."

Tony stood up. "That ain't right. Which brief is it?"

Ron knew Tony would have helped if he had been asked. "It's the 'Winning in Iraq' brief, and there's an article to go with it." Ron also knew Tony would be of very little help if he stayed.

"I don't even agree with that, much less want to touch it. Sorry, brother, but you're on your own." Tony stood up. "I'll see you tomorrow."

Ron got up and shook his hand, turned, and took his seat. He pulled open the folder and turned to his computer. He plugged away for two hours without a break. It was actually a therapy session, in a way.

When he was done, he sat back heavily in his chair and looked around. Ron Benson was alone again. He saw that the spaces that made up his office were empty of life. In the silence of his cubicle, the reality of his situation set in. The one chance he had to get promoted was gone.

Ron decided no more seeds needed to be planted and shut down his computer. He quickly changed into his civilian clothes and left the building. A one-time pass-over for promiton should be at the top of the list. A two-time pass-over is usually a shoe-in to get picked up for the next higher pay grade. For Ron, even after three times being passed over, he still was busting his ass. Why hadn't the promotion board seen that he was still a good officer? Certainly his records showed he was good enough to be a Lieutenant Colonel.

There were so many negative thoughts going through his mind. What had he done to fail? Was his work weak? Had he taken too much time off when his wife decided to move out with their daughter?

He had hoped with all his might that this year would be different. All the hope of getting out of the Pentagon and being with soldiers again was gone. Without the promotion, he was destined to remain a lowly staff officer until he was allowed to retire. Hope was no longer his companion. Hope had left the Pentagon without Major Ron Benson that night. Despair had taken her place.

CHAPTER 3

20 SEPTEMBER
O'REILLY'S TAVERN
ALEXANDRIA, VIRGINIA

Ron sat alone at the bar. It was only eight o'clock, and he had just finished his first beer. He raised his hand to get the bartenders attention so he could order another. He looked around to see what kind of crowd was escaping the chilly fall air. He noted the young couple, possibly newlyweds, fresh from their hotel and taking a break from wild, unrepressed carnal knowledge. A crowd of sorority girls screamed loudly with joy indicating more alcohol would be consumed with reckless abandon. Ron offered a silent toast to the young ladies and their good cheer. He would have given anything to feel that happy.

Three smaller groups of middle-class men, probably veterans, were solving all the world's problems before the pitcher of beer was finished. And the occasional single, much like himself, was drinking in silence. Ron wondered if the other singles in the bar had personal reasons for achieving inebriation that were better than his own. Ron was already finished with his second beer and scolded himself for drinking to fast. But that didn't tell the bartender 'no' when a third beer showed up. He took a drink from the new beer and put his back to the bar.

He was about halfway through his third Corona with lime when the two guys down the bar from him started to get louder. Ron raised his eyes from his beer and turned a sour glance at the two men. It was obvious to Ron the men were regurgitating nightly news blather about the Iraq war. They were merely average American citizens expressing what most average American citizens felt about Iraq. Ron listened silently and bit his tongue.

As much as Ron hated the situation in Iraq, he felt there was something to be said for the fact no more planes had flown into the Pentagon, or a stadium of people.

"The Son of a Bitch lied! You know it, I know it and the American people know it!" Ron thought maybe Bob Dole had turned into a drunken Democrat and had decided to come to O'Reilly's to get shit faced. "He said things and made promises! Now look at the country! We're still in Iraq, we're still losing a dozen kids a day. The media kisses his ass, and he expects us to believe him! That's total bullshit!"

Ron turned around to the bar, finished his beer and waived the bartender down to order another round. He had wished that he had kept his uniform on. It wouldn't have kept the drunk from mouthing off, but he would have been quieter. Ron was very knowledgeable of politics in Washington. He had been taught the importance of the military-political relationship. Yet he was no fan of politicians, those in congress or on a barstool. The soldier that he was, allowed him to be ready for a fight. Be it in combat in Baghdad, or a bar downtown Alexandria.

Then the second man started up. "You won't have to worry. He's a lame duck and congress has him by the balls. He won't be able to fuck up the country, because we have enough congressmen with balls to keep him in check."

Ron lost his bearing. He looked at the pair and grunted loud enough for them to hear. The pair went silent and looked his way. Their conversation suddenly included him.

"You say something friend?"

Ron looked at the men and said flatly, "I didn't say a word, pal."

The two men stepped closer. Not looking for a confrontation, but in a manner that suggested they wanted to engage in more dialogue. "You said something, buddy. I'm just curious what it was?"

Ron took a swig of his beer and said, "Actually, I exhaled in disgust."

"Why is that?"

He turned and faced the two men. "Where do you want me to start?" The more inebriated one of the two pushed away from the bar and slurred, "Wherever the fuck you want to? You one of them neo-con conservative ass kissers?"

Ron took another slow pull on his beer, but kept an eye on the two men. He had no intention of getting in a fight with two fellow Americans, in a bar, over politics, in Alexandria. Something inside him kept fueling his flame. He caught himself. Believing that it was his failure to be promoted that was making his temper appear. He took a deep breath and tried to be rational.

"First of all, I'm no neo-con. I'm not even a Republican. Secondly, I think you've been watching too much reality TV, and thirdly, I may not think much of the administration, but the President is a good man that is trying to make the world a better place." He grabbed his beer and took another pull. "Finally, I'm a Major in the United States Army, and quite frankly, I don't give a shit if there was WMD in Iraq or not, my job is to see that the people that want to destroy my country are killed." He tilted his head and added, "That means foreign or domestic!" He tensed in preparation for a punch that didn't come.

Out of the corner of his eye, Ron noticed another man had appeared. He was not with the other men, but he was intent on listening to the now heated conversation. Ron could see he was a bigger guy, perhaps six foot two, 220 pounds and pushing 45 or 50 years old. For his age, he appeared to be in pretty good shape.

"What does that mean? You think we're a threat to the country because we're anti-war?"

Ron was quick to reply, "For a drunk, you're still pretty sharp. I think you're a threat to the country because you're anti-balls, pal! I think you're scared that terrorists are going to blow the country apart and Bush is going to somehow cause a terrorist's to pop up on every corner! Instead of making them dead, as they should be. I think if you were forced to choose between Bush and radical Islam, you'd choose the Jihadis because you're afraid to fight them."

Ron took another drink. His blood was running hot. "Bush ain't afraid of 'em. He's takin' the fight to them. They're afraid of him or they'd be runnin' through our streets today. That's why you don't like him. Because he scares you."

The loud drunk could take no more. "I'm not scared of you asshole!" He took a step closer.

The onlooker stepped in between the potential combatants and said, "Gentlemen, allow me to introduce myself. Clyde is the name. Clyde Simpkins." He stuck his hand out toward the two drunks.

"We have no problem with you, buddy. Beat it."

"I know you don't, but you don't really have a problem with my friend here either." He turned to Ron, winked and said, "What's your name?"

The two drunks looked at each other and then at Clyde. Ron put his beer down and said, "Ron Benson, Mr. Simpkins. Pleased to meet you."

Clyde turned to the two drunks. "Maybe you two should go back over there, finish your beer and leave, okay?"

For just an instant, Ron thought the situation was going to end right there. The two drunks looked at each other then at Clyde. The drunk closest to the bar raised his hand as if to throw a punch at Clyde.

Like lighting, Clyde kicked out and nailed the first drunk square in the nuts! He fell down hard against the bar. Without a second thought, Clyde punched the second man in hard in the nose sending him backwards in a heap. He took two steps toward the second man, fist cocked and ready for more. Drunk two was already out cold.

Clyde shook his head and announced loudly, "Sorry folks. I apologize for this, but these two 'gentlemen' had exceeded the limit on alcohol. Please feel free to continue your evening." He turned and looked at the bartender. Ron had not noticed that the bartender had produced a huge baseball bat. He smiled at Clyde.

"I'm sorry, Danny. This was going to get ugly if I let it go." He hollered to the bouncer that had just stuck his head inside. "Lawrence! Be a good guy and see these two fellows out, please."

"Yes, Sir, Mr. Simpkins!" answered the bouncer as he hustled in to pick up the two sprawled out men.

Clyde Simpkins turned to Ron Benson and said, "Perhaps I can buy you a drink?" Not knowing any better at the time, Ron could have made the easy choice. To decline the drink and head home would have been the smart thing to do. He had just enough beer in him to impair his military decision making process.

All he had wanted was to drowned his frustration with his job. Ron wanted to achieve a little normalcy in his life. But he was depressed, he was angry and he was alone. He had nearly gotten into a fight for no good reason. Inside he knew the best thing that could happen to him was a change. He just didn't know what that change was.

Clyde Simpkins had essentially rescued him from an ugly situation. The unforeseen rescue had pulled Ron Benson from the bowels of despair into uncharted territory. To a path where there would be no return to normalcy.

They sat in a booth in the back of 'O'Reilly's'. Clyde Simpkins put some ice in a plastic bag the server gave him and laid it on top of his bruised hand. Ron looked at him cautiously.

Ron noticed Clyde had gel in his graying hair, which was combed straight back and accentuated his sharp facial features. Ron thought he looked like an Italian hit man or a PI from a fifties film. He talked fast, as if he was from the northeast, perhaps New York City. His jacket and tie matched, but didn't fit right or something. Ron couldn't put his finger on it. Maybe he was in a hurry when he got dressed.

What he could tell, was that as weird looking as Clyde was, he was very confident. Confident, friendly and in shape. Like a used car salesman but without anything to sell. The man was older than Ron, but lean. There was no little gut or love handles around Clyde's waist. Actually, Clyde was quite physically gifted. It was no wonder he kicked the crap out of the two drunks. Clyde was in damn good shape.

For some reason, Clyde's appearance was no matter to Ron. As strangely dressed as he was, Clyde had a demeanor, a certain calming presence that Ron felt comfortable with. It wasn't easy for Ron to relax around anyone, especially strangers. Clyde was different. As strange as the first meeting was, it seemed Ron Benson had known Clyde his whole life.

Two more drinks came. Ron was struggling to put his work behind him. Clyde was living in the present and discussed the cause of the argument. "I only heard a little bit of the discussion there, but I'm guessing you're a fan of Bush's?"

Ron nodded. "Not of any of the rest of the clowns, but I have respect for what he's trying to accomplish."

"What about Rummy?"

Ron eyed Clyde Simpkins carefully. Clyde was not the enemy, but this was a touchy subject and dangerous ground for the Major. "I do not particularly care for the Secretary."

Clyde laughed out loud and smacked the table playfully. "I knew it! I knew it!" Ron smiled. "You're a military man, aren't you?" asked Clyde. Ron nodded. "Let me guess! Air Force?"

"Army."

"Damn! Army? I should have let you take those two dirt bags on by yourself."

Ron shook his head no. "I appreciate the help." Then he thought of something. "How come the bartender and the bouncer didn't throw you, and me for that matter, out as well?"

"Oh, we have an understanding."

Ron nodded again. He rolled his hand as if to say, 'little more info please'.

Clyde leaned forward and quietly said, "Sometimes I work here."

That seemed normal enough. Ron wondered why he leaned so close and seemed to think it was a secret.

Clyde changed the subject. "So are you still in the Army?"

Ron's smile was subdued. "Yeah. Still. For awhile. I'm trying to make it two more years so I can get my twenty."

"What rank are you?"

"Just a major." Ron took a drink and said, "May finish that way if I'm lucky."

"A Major! Wow! That ain't too shabby!" It surprised Ron that Clyde seemed to know something about the military. After watching the speed with which Clyde dispatched the drunks, Ron wondered if Clyde was not a military man himself.

Clyde answered the question before it was asked. "United States Marines Corps! Semper Fi!" He held up a quick, yet professional looking salute. At that instant, it dawned on Ron that the nerdy image presented was all an act. Clyde Simpkins at one time was probably one helluva good Marine.

"I had to give up the Corps about twenty years ago. I was a Corporal, almost ready for stripes. Saw a little bit of action in Greneda." He looked over to a wall at the clock and back at Ron. "Came back and decided I had different priorities."

"Go to college?"

"Nope. Dishonorable discharge for selling pot!"

Ron coughed into his beer and then laughed out loud. "Damn! I shouldn't even be seen with you!"

"Too late now, brother! The bars security cameras already have us sitting together on tape. As far as anyone would know, we're best buddies."

Ron noticed the camera's around the bar. "You're a regular here aren't you?"

"I've seen you in here a time or two." The comment surprised Ron. He was unaware anyone had noiced him in the tavern.

"Yup." He nervously looked at his watch. "You're okay, Major Benson. I see you're wearing a ring. Shouldn't you be home right now?"

It was right then that Ron realized Clyde was very observant. Ron wasn't certain that Clyde's question was an innocent one. He seemed more like an intelligence officer. One that already had the answers to his questions. Ron looked down at the ring and turned it gingerly with his free hand. "No. I'm divorced."

"I see. Divorced." The answer was no surprise to Clyde. "How long now?"

Ron didn't have to do the math. "Two years." Clyde nodded and looked down at the ring and watched Ron continue to turn it slowly. "Two years, four months and three days now."

Clyde sat back in the booth. Look directly into Ron Bensons' eyes and said, "You need to get laid."

Ron turned his attention to his beer and tried to think of a witty response. When nothing came, he settled for an honest question. "What makes you say that?"

"Let's see. First of all, you got real upset with those two chumps about nothing. You were ready to fight them, when you're actually mad about something else. Your ex-wife maybe?" Ron didn't answer. He looked down at his drink.

"Then you tell me you're a Major, after I practically force it out of you! So you aren't proud of your service or the fact that you're only a Major? I think you said, 'May finish that way'?" Ron looked up and flushed red in the face. Clyde smiled and nodded. He knew he was on the mark.

"And finally, you're divorced and you still wear the ring! Plus, you've been divorced for over two years." Ron looked down again. "I bet you haven't been laid since you split up, right?"

Ron started to lie, but couldn't. "Once, right after the divorce was final."

"Was weird wasn't it?"

"Yeah, it was weird."

"You ready to try again?"

"What?"

Clyde knew Ron had heard the question. He leaned forward again. Ron sat back and said, "Look Clyde, I'm not gay if that's where you're heading?"

Clyde laughed. "Listen. I could use a little help tonight." Ron was confused. "I'm kind of," he looked for the words. "I'm double booked and I could use some help."

Ron wasn't exactly sure what Clyde meant. He thought maybe he had drunk too many beers. "Help?"

Clyde exhaled loudly. "You're gonna make me spell this out aren't you?" Ron shrugged his shoulders.

"Look, Major. Help me out. I helped you out earlier. Can you do this for me? It's probably the best thing for both of us?" He was nodding when he said it.

"What do you want me to do?" "I've got a couple girls coming and both of them are looking for some action tonight."

"Action? You just got into a fight? What else do you have in mind?"

Clyde titled his head and said slowly, "Action. As in sex, Major Benson." Ron was still confused. "Now the blond one, Leslie, she's mine tonight. But she has this friend, kind of a librarian type. She wears these glasses that are kind of funky, but she has a tight body that you'll just . . ."

"Wait, wait, wait a second here!" Ron took a deep breath and looked around to see if anyone was listening. "How can you be so sure they want . . . you know. Sex tonight! I mean, we haven't met. This little 'librarian' you called her? Suppose she doesn't like me? Maybe I could take her to a movie or something?"

Clyde put his hand up. "Major Benson." Ron stopped talking, but clearly had questions. Clyde thought he looked scared. To Clyde it seemed like Ron might actually get up and run away. Yet he knew the Major better than Ron knew himself. He understood Ron wanted something different in his life from what the Pentagon was giving him. Major Benson was at a fine line in his life. Clyde could either push him back to the Pentagon, or pull him across into a world he needed to enter. "Ron. My brother!" He stuck his hands out. "Have I steered you wrong at all tonight?" Clyde noticed Ron shake his head. "Have I lied to you?" Again, no. "Can you please screw this woman for me tonight?"

Ron hesitated. "How can you be so certain that she . . ."

Again the hands came up.

Ron suddenly had a terrible thought. "She's the ugly one, isn't she?"

It was Clyde's turn to laugh. "I told you she is tight. Not beautiful, but very attractive. You'll like her and she'll absolutely love you."

Ron sat back and finished his beer. Up to this point, Clyde had not lied to Ron. If he said the woman was attractive, it was probably the truth. His mind raced back to how he had wished for the immediate tryst with Mary back at the office. Sex would be a great way to forget about how shitty he felt. *Hell, if I have any opportunity for sex, I should jump at the chance.*

Just as quickly as the thought of coitus entered his mind, another terrible thought appeared. *Could I even do it? It had been so long? Crap. How many beers did I drink? Could he still get it up?* He silently fought the urge to try to stimulate himself to see if he could get a hard on.

Clyde knew exactly what he was doing. "You'll be fine. It's like riding a bike, without the ass pain." He looked at his watch.

Ron checked his. Nine o'clock.

Clyde smiled and looked across the room. "Right on time." He looked at Ron and said, "I love punctual clients. Smile. Here they come."

Ron smiled until the word sank in. "Clients?" *What the hell does that mean?*

Ron turned and followed Clyde's gaze. Two women were walking across the bar towards them. Ron watched the tall blond stroll across the floor, her high heels clicking daintily on the bar room floor. A low-cut red dress that made her hard to miss, Ron couldn't help but notice all the men in the bar turn their heads to catch a glimpse of the Venus that had just entered the room. She was just plain beautiful.

The second woman, following two steps behind was much smaller, maybe five foot two. She was a brunette and wore her hair up. She had on black rim glasses and Ron knew instantly why Clyde had called her a 'librarian'. She wore a sweater over a turtleneck and had a shirt that went all the way down to her knees.

Clyde turned to Ron and handed him a business card and two condoms. Then he leaned close to Ron and said, "Your name is Steve, okay?"

Ron didn't bother to look at the card, but knew instantly what the condoms were. He quickly tucked the items into his pocket and nodded in agreement. "Sure."

"Hello, ladies! You two look wonderful tonight! How are you?" He stood up and greeted them both with a kiss on the cheek. Ron wasn't sure, but he thought Clyde had grabbed the blonde's ass and gave it a squeeze.

"I'd like you to meet my friend, Steve. Steve, this is Leslie." Ron stood and shook the tall blonde's hand and she smiled gleefully. She was actually an inch taller than Ron. Then Clyde turned to brunette and said, "This is Edie. Edie, this is my friend, Steve."

Steve shook Edie's hand and to his surprise, it was warm and soft. He held her hand just a little longer than he intended, but she didn't pull away. He looked into her glasses and saw two friendly eyes looking at him curiously. He pulled his hand back and offered, "Would you like to sit here?"

Edie smiled shyly, moved slowly by Ron and sat in the booth. He couldn't help by think she was a very pretty woman as he slid in the booth next to her. He made a point not to get to close at first, less he seem too forward.

An hour later, after returning from the rest room, Edie had removed the sweater. Ron couldn't help but notice she filled out the turtleneck fantastically. He slid in next to her, but close enough to touch her this time.

Soon after ten o'clock, Leslie checked her watch and said rather nonchalantly. "Look, I've got a big day at work tomorrow. Clyde, would it be all right if we got done early?"

"If it's all right with Edie?" He looked across the table and stuck out his hand toward Edie. "Edie, Sweetie. I need to devote some special attention to Leslie tonight. Is it okay with you if Steve sat in for me?"

Edie, holding Clyde's hand turned and looked at Ron. "I don't know Clyde. You're the best and you know I'm worried about being with other guys."

"Honey, Steve here is brand new to this. He is clean, he is ready and he needs you." He squeezed her hand. "I'll give you fifty percent off too." She looked over at Ron who was now looking at Clyde confused again.

"If he's not any good, I get a rain check, right?"

Clyde pulled her hand to his lips, kissed it and said, "Sweetie, if you don't love this man, I'll see you tomorrow!"

Edie bit her lip and said, "That sounds fair enough."

Clyde nearly pushed Leslie out of the booth. "You two have fun." He stuck out his hand and winked at Ron. "Steve, you be gentle with her or she'll hurt you!" He looked at Edie. "Don't worry about the discount; just give him what he's worth!" Steve and Leslie nearly sprinted out the door.

Edie seemed pleased with the last comment. She looked over the top of her glasses at 'Steve' and said, "I kind of have a long day tomorrow too. Can we leave now?"

Ron stood and yelled across the bar, "Check, please!"

CHAPTER 4

20 SEPTEMBER
APARTMENT 7 C
4307 VENTNOR AVENUE
ALEXANDRIA, VIRGINIA

Edie led Ron into the dark hallway and turned on the light. It took a moment for Ron's eyes to adjust. They were in a small living room. There was a large couch that dominated the room, a lounge chair and a lamp. Across from the couch was a large 35-inch television stacked on some cinder blocks and boards against the far wall.

Edie quickly went down another hallway and said over her shoulder, "You can drop your jacket on the couch. I'd offer you a drink, but I'd rather not waste the time if it's okay with you?"

Ron nodded to no one else in the room. "Sure." Then he hollered towards the hallway. "That's fine!"

She quickly came back in the room wearing only a black bra, an unbuttoned silk shirt and her stockings. That was it. Ron started to say something, but she put a hand over his mouth. "Shhhhh! Don't wake up the neighbors, okay?"

Ron nodded again. She turned on the lamp then hit a wall switch that turned off the overhead light. Then she stopped and looked at Ron. There was an awkward pause and Edie suddenly snapped her fingers. "I noticed you're wearing a wedding ring. You're not married, are you?"

Ron was still trying to pry his eyes away from her body when the question finally hit him. He looked down at the ring. "Um, I'm divorced. I just wear it out of habit." He started to bite his tongue. That didn't sound like a very good reason to keep a wedding ring on. The trouble was, it was the truth.

She nodded and stepped in front of the couch. The answer was good enough for her. "Well, come on, Steve! You're looking at me and I'm not seein' you! You won't get a tip if you don't move a little faster. I don't have all night!"

Ron was shocked. He had expected something different. Maybe, some kind of candles, or a bottle of wine? Something other than being immediately drawn into a business transaction?

He stuttered, but the word, "Sure," came out. He exhaled loudly and started taking off his clothes. Edie for her part had taken off the black rimmed glasses. She reached up and pulled something from behind her head. Her hair came down around her and she shook her head. Right before his eyes, the librarian was gone, replaced by a fantastically beautiful young woman.

Ron need not have worried about getting an erection. He had one before he got his shoes off. He nearly fell over taking off his pants as he started to hurry. It made Edie laugh and she sat on the couch. "We could go into the bedroom, but it's a mess and the walls are thin."

Ron moved closer to her. "This is fine." He sat next to her in just his Hanes briefs.

She looked him up and down and said, "Yes." She reached over and slowly put her hand in his Hanes. "Yes, it is."

Somehow, someway, the stars above came into alignment and Ron managed to 'survive' for fifteen minutes.

He was still slightly out of breath when he pushed himself off the floor and rolled over to lean against the couch. He exhaled heavily and said, "I would have liked to . . ." He looked for the words but she already knew what he was going to say.

"Liked to have lasted longer?"

He nodded and looked down at his naked, sweaty body. "Yeah."

She picked herself up off the floor and leaned next to the couch close to him. "Trust me. You did fantastic." She reached a hand over and rested it on his bare chest. She rubbed it softly.

Ron felt the stirrings in his loins again. He shook his head slowly and smiled. *Was it possible?* Could he actually even consider a second encounter?

Again, Edie must have read his mind. Even in the dim light, she could see he was beginning to rise to the occasion. She playfully smacked his chest. "Don't even think about it, buddy!"

She got up off the floor and found her shirt and put it on. She didn't bother with anything else. "Look, I really do have a busy day tomorrow. I've got to retype about a dozen files. I hate to do this to you, but I need to go to bed."

For just an instant, Ron was hurt. "I'm just guessing that means alone?"

"Alone."

He nodded and stood up. Somewhat proud that his manhood was semi-aroused again. He found his briefs and put them on. It took him a moment to locate all his clothes as they were in several locations around the living room.

Edie walked over to her purse and pulled out a bill.

"I think you deserve a tip for the exceptional effort. It wasn't a full hour, but if I didn't need to get to bed, I make you work for this." She handed him the bill, reached over and kissed his cheek.

As she pulled away, she looked up at him. "I usually just call Clyde when I'm horny. Would it be all right if I called you?"

Ron looked down at her. It took every ounce of energy he had to keep form yelling, "HELL YEAH!" He had to think quickly. "It's probably best if you just called Clyde for now." He could tell she was disappointed.

"I'm sorry you didn't get a full hour."

"Don't worry about it! You're getting paid for the event, not the clock." Ron thought about what she said and it made sense.

He stepped close to her, looked down at her beautiful face and kissed her softly. "Thanks, Edie. That's the best advice I've heard in years." He pulled back to look at her. "You need to go to bed. I hope I'll see you again." Her kissed her forehead and finished getting dressed.

She saw him out the door, got another kiss goodbye then quietly shut the door.

Ron Benson stood in the dimly lit hallway. It had been an interesting evening. What started out with anger and despair had turned in to something completely different. He wasn't sure how many laws he had broken or what they were. He just knew that it didn't matter. He felt better than he'd felt months. Better than he had in at least two years, four months and four days.

CHAPTER 5

21 SEPTEMBER
313 WASHINGTON AVE
APARTMENT 24
FAIRFAX, VIRGINIA

Ron opened the door to his tiny one bedroom apartment just after one o'clock. He quickly went over to check the answering machine. He threw his bags on the couch and checked the messages. The first two were from credit card companies and Ron made a mental note to get on the 'no call' list. The third call was from his daughter, Ally.

"Hi, Dad! Just wanted to let you know the great news. I got accepted to Princeton!" Ron was ecstatic. "I wanted to tell you as soon as I found out. I'll talk to you tomorrow. Love you!"

He turned off the machine. "Princeton! Wow!" He was still smiling after he got a glass of water and headed to the bathroom to take a quick shower. The remnants of the previous hour's events were washed away, but not the memory. As he dried off, he found himself thinking about Edie. How sex with her had been so casual. So business like.

He headed to bed and sat down heavily. He looked at the dresser and grabbed his wallet. He pulled out the bill Edie had given him and the card from Clyde. He lay back against his pillow and examined both. The card read:

Clyde "the Ride" Simpkins
1-(888) 555-8141

"The Ride" being printed on the business card gave Ron a chuckle. Clyde was a piece of work. He looked at the back and it was blank. That was all that was on the card. He unfolded the money and saw a rather new one hundred dollar bill. Ron smiled again. He had just been paid to have sex.

Then he noticed the wedding ring. His mind drifted to his ex-wife, Sandra. He was angry at himself for thinking she would give a damn what he did. Then he thought of Ally. She could be attending one of the best, most expensive colleges in America. Was he going to deprive her of that opportunity because of a lack of money?

For eighteen year's he had been disciplined to make decisions based on the facts. He considered the pros and the cons of what he was doing. On one hand, he had no idea how many laws of the Uniformed Code of Military Justice he had broken. On the other, he had just made the easiest hundred dollars he'd ever made in his life. He looked at the picture of Sandra on his bedside table. No one knew what he had done. No one would. As far as his peers knew, he was a monk. He had no social life.

He took one more look at the business card. He placed Clyde's card back in his wallet, along with the hundred-dollar bill. He looked at the ring on his finger and slowly pulled it off. He put it in the drawer in his nightstand.

He rolled over and turned off the light. He thought about what he had to do at the office. He had to give a draft presentation to H2 No, but it just didn't seem to matter. The Colonel would make some cosmetic changes. Hell, his military career was all but over anyway. He quickly dismissed the thought of power point slides and his boss.

Ron looked over at the pictures on his nightstand. One was of Sandy with a big smile on her face. It was taken over ten years ago, back when she was much happier with Ron. The other picture was of Ally. It was taken only four months ago. He knew what Sandy would think of his actions with Edie. He wasn't sure he would be able to explain them to Ally. But his knew line of work was something his daughter never needed to know about. It would probably hurt him more than her if she knew what he was thinking about doing. But if he was careful, she would never find out.

At that moment, it really wasn't too tough a decision to make. He loved his family, but they weren't there. He loved the military and she too had abandoned Ron. He didn't need the damn military decision making process to make up his mind. He made a mental note to call Clyde first thing in the morning.

CHAPTER 6

21 SEPTEMBER
C RING - 117
PENTAGON, WASHINGTON DC.

Christopher Watson was a Department of Defense contractor hired to provide Public Affairs mentoring to the junior officers in the Public Affairs Office. Before he was the "Yoda" of the PA Office, Chris Watson was an Armor Officer in the United States Army. He was 54 years, balding, with ever present bi-focals, Chris stood hunched over a walking cane and talked very slowly. He had served as a Brigade Commander in the First Infantry Division in the Gulf War. His Texas accent covered an intellect second to none. He wasn't called 'Yoda' for nothing. Behind the spectacles and the few remaining gray hairs, was one of the sharpest Public Affairs minds in America.

Whenever Ron saw 'Yoda' he smiled, because he knew he was about to learn something. Ron's aviation experience had been a great base for learning about the Army and warfare, but he didn't know enough about Public Affairs or Diplomacy to be the best at his current job. Yoda was the one person in the office that taught him how to do his job. Ron had learned nearly as much as he could from Chris. Recently, the lessons were coming fewer and Yoda was being directed to assign little missions, essentially tasks, that H2 No wanted to be done but didn't care to task him self to be done. These little tasks, became known around the office as "drive-by taskers". Yoda was getting a bad reputation as the driver of the drive-by's.

When Ron walked into his cubicle and saw Chris Watson there, he knew from Chris's smug expression that he was in the driver's seat. Another tasker from H2 No was waiting for him.

He knew it wasn't Yoda's fault, so he tried to handle the imminent ass pain the best he could. "Good morning, Chris. How's it going?"

"Going just fine for me, Major Benson." The tone, combined with Yoda's smile told the tale too soon. He had to throw another jab at Ron. "I hope your day was going well." The southern accent didn't disguise the fact that Ron Benson was about to have an SUV dump crap in his lap at 80 miles per hour.

Ron put his coffee down and sat heavily in his chair. "What's the deal, Chris?"

Chris Watson walked in and sat in Tony Parkman's chair as if it was his own. "I fixed the F22 briefing. The boss thinks you actually love the airplane now."

Ron was instantly pissed at Chris for changing his work. "That's bullshit, Chris! You need to tell him the truth! DOD shouldn't be buying that money pit."

Years of wisdom were displayed in the smooth, slow southern accent. "That's not your call, young Major."

"Then you need to tell him that's your work, because I think everyone of those planes should be scraped. They aren't needed in the next decade."

Chris looked over his glasses and spoke slowly. "Powers that be want the airplane, so it's gonna happen." Ron shook his head and sat back in his chair. He knew the lesson was only part of the reason why Chris was there.

Chris turned to the real reason he was there. "There's an Inspector General team coming to visit us. It appears that one of the reasons the IG team is coming through is to see what the PAO relationship is to Information Operations. Do you know what IO is?"

"I know it's oriented on operations more than intelligence." Ron thought a bit longer and saw Yoda tilt his head and smile. "I also seem to remember the General got into a bit of a pissin' contest over what its real role is." Major General Patrick McCoy was the military head of Public Affairs. He was no fan of IO and after he told his superiors that fact, he was rumored to have been told to shut the hell up and give the storyline. That or find a good retirement area.

Chris Watson was smiling. "Exactly. What is the role of IO and how does it relate to the PAO?"

It was Ron's turn to think. "Some people think there is no relationship at all. But some of the rub was the fact that Public Affairs actually support's Information Operations roll in warfighting."

"Now why would that tick people off, Major Benson?"

Ron Benson remembered Yoda's first lesson about the Pentagon. If you wanted to figure something out, "follow the money". Ron leaned forward in the chair and talked lower. "You're the one that always told me that this place runs on money." He saw Chris smile. "If IO has its own lane, it will get fully funded and resourced."

"And . . ."

Ron absently scratched his head with his pencil. "If it gets funded and resourced, some other department . . . " Ron was thinking out loud.

"If it gets funding, some one has to pay for it."

Ron nodded. "Some other area or lane loses funding." He understood what Chris was saying, but wasn't sure what the task was.

Chris didn't keep him in suspense. "Colonel Henderson wants you to write a talking point paper on the value and benefit of PAO as it relates to IO. You are to emphasize how wonderful we are and how we have, nor want, anything to do with information operations."

"How long and what's the suspense?"

"Five pages by close of business."

Ron shook his head in disgust. He didn't really understand IO, so he'd have to research it some before he got on the paper. "It might take me until tomorrow."

Chris nodded and stood up. "If he needed it by close of business, first thing tomorrow should be just fine. I'll let the Colonel know it'll be on his desk tomorrow, okay?"

Ron nodded. "First thing, Chris."

Watson smiled and turned to walk away.

Ron called out, "Chris, what if I find out we should be supporting this IO branch?"

Yoda looked over his glasses and offered a knowing smile. "Major Benson, I recommend you don't find anything that indicates that result."

Ron did a quick online search about Info Ops and found a contact in the J3 operations office that he arranged to meet in the cafeteria at 10 o'clock. On his way to the cafeteria, he stepped outside to make a quick phone call to Clyde.

Unfortunately, all he got was an answering machine. He left his name and number and asked if he could call him back. As he hung up he realized that he had left the message with genuine sincerity. He really did hope Clyde Simpkins would call him back. With luck, he could have the paper finished by seven and he could meet Clyde that night.

At the cafeteria, Ron met Major David Meyerson. Dave Meyerson was a US Army Psychological Operations officer. He was assigned to the Info Ops shop in the Pentagon. He had been there for over a year and was well aware of all the politics going on between IO and PAO. He also knew that the Psychological Operations community was knee deep in the potential loss of money to IO as well as Public Affairs.

"I can tell you everything you want to know about IO. As far as PAO, that's your department." Ron nodded in agreement. He pulled out a small notepad and began to take notes.

Dave Meyerson told Ron that IO was actually a combination of Psychological Operations, Computer Network Operations, Operational Security and Electronic Warfare. There were "other things" that Dave couldn't go in to specifically. But all the elements of IO were operationally focused actions that supported military commanders at every geographic command. Then he added that all the geographic commands were being forced to create IO Divisions from within their own staffs.

"IO is a tremendous way to coordinate and de-conflict information that supports the warfighters at the front. The problem is, nobody has resources to pay for it."

Ron nodded his understanding. Chris had always told him that money was what drove decisions in the Pentagon. The decisions being made about Info Ops was no different. "So what is the value added?"

"The value added is it breaks down walls between the services, who are the force providers, and forces information sharing. It allows for coordination of inter-service and even interagency activity with other branches of government like State and Commerce."

Ron shook his head. "I can see all kinds of problems. The services don't want to let each other know what they know or what they are doing. The other branches don't want any more funds to go to DOD. If it has to be resourced with funding and personnel, then nobody wants to pay for it." That led him to another question. "What is the current role of PA in IO?"

"Right now, it supports IO, but only with information dissemination. Rumor is they are looking to make PAO a partial bill payer." Dave Meyerson shook his head. "Probably your boss already knows that."

Ron nodded. "He's about to know more about IO than he ever wanted. Thanks, Dave."

Dave wasn't done. "One other thing."

"What's that?"

"I wouldn't count on it going away. IO is here to stay, so you PA guys are going to need to figure out how to play nice." Ron nodded. He understood the ramifications for Public Affairs.

The whole concept was frustrating to Ron. How was this really going to help an 18 year-old private in Baghdad that was getting shot at. The truth was that kid needed ammo and a target. Of course that was too simplistic. The enemy was using the internet as Information Operations and kicking ass with video speeches on Al Jazeera.

That's when it hit him. He understood the whole bureaucratic nightmare of the rice bowl factory that was the Pentagon. The enemy didn't have to staff his video's or get approval from his chain of command. He gave an passionate speech in front of a camera and people watching new what his message was. He would kill his enemy or die trying. He respected his enemy for that.

But it offered him a moment of clarity. The Pentagon was no longer serious about killing enemies. They had joined the political chorus that called for diplomacy and dialogue to defeat the enemy. This enemy would not be defeated by words. It was at that moment that he decided he needed a new line of work.

The interview left Ron in terrific spirits. He now had purpose. The priority was to take care of his current task. He had the outline already in his head as he left the cafeteria. The paper would not be on the value of PA or how it relates to IO. The paper would say PA needed to support IO. Anything that helped the troops forward as much as Info Ops needed all the help it could get. If that meant taking funding from PA, so be it. *Henderson was gonna shit.* Ron smiled at the thought. He didn't care about Henderson right then. His mind was already miles outside the Pentagon.

CHAPTER 7

21 SEPTEMBER
"OLD TOWN TAVERN"
GEORGETOWN, VIRGINIA

"I have to admit I was surprised you called. And so soon!" Clyde was slowly peeling the label off an ice cold Amstel Light.

"I'm kind of surprised myself." He exhaled slowly. "I really had fun last night. Edie was fun. Hell making money for . . ."

Clyde broke into an old Elton John song. "Gettin' paid for bein' laid! Yes, that's the name of the game!"

Ron smiled. "I suppose the reason I called is that," he paused a second, nodded as if to confirm his decision and said, "I want to do this. I liked it."

Clyde stopped peeling his label. "You sure you want to? I mean. You've got 18 years in the Army? You're violating at least one law of the Uniformed Code of Military Justice. It's not too late to just forget what happened. Chalk it up to one of those life experiences."

Ron had decided. "I can honestly tell you, I need to do this. I have a shit job, for a shit boss. I work at the Pentagon for Christ sake! This is a chance for me to escape. To find out I have a life to live. I have been so absorbed in my work, that I have missed some of my life."

Clyde nodded. "What about your wife?" He noticed the ring was gone. Clyde wanted to make certain that Ron wouldn't have any more guilt over his failed marriage. The ring was on his finger for a long time after the marriage was over.

Ron thought about Sandy Benson. She knew all too well that Ron was totally dedicated to the Army. It had cost him her love. For the first time, he realized that was the reason she was his ex.

"Clyde, when we were married, I was a dedicated fool. I forced her to change. I wanted her to be the perfect Army wife. Thinking that would help my career.

"When we were at Ft. Bragg, I saw all these trophy wives running around. You know what I mean?" Clyde nodded and cracked a small smile. He understood exactly what Ron meant. "I mean half of 'em were Barbie dolls, the other half would do anything to see there husbands succeed."

"I've seen it before."

Ron wasn't done. "Sandy was a little overweight after having our daughter. She never really lost the weight. After I got promoted to Major, I asked her to have liposuction." He shook his head. The weight of reality was being lifted as he spoke. "I was so obsessed with my career that I wanted my wife to risk surgery. Thinking that would get me promoted. How screwed up is that?"

Clyde was no longer smiling. "That why she left?"

"That and my working 15 or 16 hours a day." Ron took a drink of his beer. "She left when I deployed to Kosovo in '99. I don't blame her. It's kind of tough on our daughter. They're getting alimony and she has a live in now that won't marry her. I guess she's happier with him. So I'm happy for her. I don't care for the asshole she's with one bit.

"So my life has been the army for the last six years. Knowing what I know now, my life has been shit for all that time." He shook his head and looked at Clyde. "I've been missing too many nights like last night, for too damn long."

Clyde was frowning now. "First thing brother, you need to know, is get off the pity. Quit feeling so freaking sorry for yourself." Ron hadn't quite expected that response.

Clyde continued, "You've had a great military career so far. You're a Major in the best Army on earth. You're healthy. Even though your family isn't with you, you have one. I bet they still love you too." The truth was sinking in and Ron nodded. "You have absolutely nothing to complain about."

"That little girl I love wants to go to college. I'm complaining because I can't pay for that on a major's salary. My military career has peeked. I'm essentially alone and responsible to no one except her. And that hundred dollars Edie gave me wasn't too shabby either."

Ron looked into Clyde's eyes. "I want to do this." Clyde looked non-committal. "I need to do this."

Clyde's small smile came back. "All right, Major." He shook his head as if he wasn't so sure about Ron's decision but was still smiling. "This is against my better judgment. It just so happens that I'm getting a bit overbooked lately. I could use some help."

"Let's get some things straight. If you're gonna do this, you work for me. That means you have to listen to what I tell you, do what I tell you and be what I tell you. Is that a deal?"

Ron nodded and stuck his hand out. "That's a deal I can live with."

Clyde looked at the hand. He was slow to take it. He had never had a partner before. "I hope I don't regret this." Ron smiled broadly.

Clyde looked around to make sure no one was listening to him. "Here's Clyde's rules. Your job is to satisfy the customer. That's what all these women are to you. Customers. They are your business. If you want more business, you have to take care of your customers. That means always leave them wanting repeat business." Ron nodded. He found himself wishing he had his notepad.

"Second. Always. I mean always wear a condom. You are not just protecting yourself, you are protecting your clients." Ron understood completely and made no protest.

"Third, you think you have secrets at the Pentagon. They are nothing compared to the secrets in this line of work. If you develop a client list, you must keep your meetings, your conversations and your activities absolutely secret. Is that clear?"

"Perfectly."

Clyde was smiling broadly again. "I will also tell you something. I did get a call from Edie."

"You did?"

"Yeah. She said you were 'good'!"

"Just 'good'?"

"That's a good compliment for your first time. What was it? Nearly two years?"

Ron was visibly smiling. "Good works for me."

"She also said she like to see you again."

"I can do that! When?"

"At our prices, next week at the earliest. She's not one of our more well to do clients. But she is regular. And I'll take a good regular customer of a wealthy sometime client any day."

Ron thought of something. "That reminds me. How much do I owe you? From last night's effort."

"Let's call that one a 'gimme'. Nothing." Clyde did some mental math. "Let's see. I've never had a partner before. I guess since I'm arranging things and keeping you out of the limelight. How about a flat ten percent?"

"That's all you want?"

"You want me to charge you more?"

"Oh, no! That's fine by me! Ten percent it is."

"Glad we got that out of the way. Now let me tell you another secret." Clyde leaned forward. "You want to know what women want?"

"Doesn't every guy?"

Clyde nodded, "They want to be respected. They want to be appreciated. Not taken for granted." Ron shook his head. "They want to be treated the same way we want to be treated. Only nicer."

"I can do that."

"If they want dinner. It's dinner. If they want to dance. It's dancing."

"Dancing, eh?"

"If you don't know how, learn. You are an escort. You are a gigolo. A professional. It is your job to do what your client wants to do. Got it?"

"Roger that." Ron thought about it. There wasn't anything they discussed that he was not willing to do. In fact, it might be good for him.

"And another thing."

"Yeah?"

"You need to lose some weight."

Ron was stunned. "What?"

"You heard me. You need to lose about ten pounds. Maybe fifteen!"

"Fifteen pounds! Damn, I meet the Army standards."

"You don't meet mine."

Clyde was actually in pretty good shape. He was very lean for his 42 year-old body. "I guess I can do that." Deep down Ron knew Clyde was spot on. He did have a little gut roll starting to show. He tried to rationalize the directive thinking that he would be doing the Army a favor by getting in better shape. As if the Army would appreciate a 40 year old gigolo in its ranks. But, if his new 'job' motivated him to lose those extra pounds, that was just a bonus towards the old job.

"How much work can you handle?"

"Being how I still have my day job, I figure three, maybe four times a week?"

"We'll try three for now. When can you start?"

Ron looked at his watch and smiled broadly, "How about now?"

"How about not! It's gonna take me a day or two to get you some work."

Ron couldn't hide his disappointment, but nodded in agreement. "Okay."

"Besides, you need to get to losing that weight!"

Ron smiled again. "Don't be a jerk," he paused before adding, "Boss."

"Hey. I may be a male slut, but I have standards! And if you're working with me, you better have them too!" He raised his beer to toast. "Agreed?"

The beer bottles touched loudly. "Agreed!"

"To future success!"

Ron took a long pull on his Corona. "I guess that's the last of that for me for awhile."

"Good! I've got an appointment I need to get ready for." Clyde stuck out his hand as he got up from the table. "I'll give you a call tomorrow."

"Can't wait."

Ron sat at the table alone. He had just agreed to become a male 'escort'. A gigolo of the first order. It sounded so much better than prostitute or slut. He leaned back and looked at the surveillance camera as it watched him in silence. He wondered if it would make a difference what they called it when he was court-martialed.

CHAPTER 8

28 SEPTEMBER
C RING – 117
PENTAGON, WASHINGTON DC.

Ron couldn't help but smile. Tony Parkman was at a five-foot hover over his desk because he was so pissed off. "Why can't you do it? Shit, I don't want to talk to the IG team!"

"You know damn well, H2 No, doesn't want me talking to them. He's still pissed at me for that F22 paper." Ron was doing everything he could to suppress his smile.

Chris Watson popped around the corner and entered the cubicle. "Major Parkman, Major Benson, good day to you both."

Parkman whined, "What the hell do you want? You got us another drive by shit sandwich? You know every time we see you lately, there seems to be some kind of bullshit deal for one of us. Aren't contractors supposed to do the work?"

Chris countered, "Why, Major Parkman? Did the string on your Tampon break?" Ron laughed out loud. But, Watson wasn't done. "I haven't heard that much whining since I left the Colonel's office. Are you trying to be like your Daddy?"

"Ouch, Baby!" winced Ron.

Tony Parkman shook his head in defeat. What was friendly banter, was done. Yoda got down to business. "Major Benson, you first. Colonel Henderson was none to pleased with your paper on Info Ops. In fact, he wants you to re-write it."

Ron sat back in his chair. "Not gonna happen, Mr. Watson."

Chris pulled up a chair and sat down. He acted as if Tony Parkman was no longer there. "You don't seem to understand the situation. You're treading on thin ice. The F22 brief was finally accepted after I fixed it."

"You mean, changed it."

"Don't digress on me." Yoda leaned forward and looked over his bi-focals. "I have problems with this PA, IO relationship, just like you. Colonel Henderson doesn't see it the way we do. His view point is totally from the PA Office."

"Why don't you fix my paper, Chris? I'm not gonna change it." Ron looked at the ceiling trying to pretend he didn't care.

"Not this time, amigo. Henderson knows I worked on your brief. I have been ordered not to touch the paper. He wants you to write it showing PA in a more positive light than you did on your last effort."

Ron was tempted to go down the hall and confront Henderson. If he wanted the paper to say PA was more important than Info Ops, let him write the paper. Chris Watson could see the anger in the Majors' face.

Yoda wasn't working in PA because he was naive. He was stern when he addressed his favorite Major. "You need to think about this." He moved closer to Ron. "I know you're trying to look out for the troops, and you probably even think this paper can help do that. You should just take some time and reconsider. You can't go talk to the Colonel right now either."

Ron let Chris Watson's words sink in, but he was just agitated enough to head down the hall. A sense of better judgment kept him in his chair. "I'll think about it. But lying on this paper isn't the right thing to do."

Yoda had won again. "I'll tell the colonel you are reconsidering the wording."

It only made Ron angrier. "I know you're trying to help, Chris. You can tell him anything you want. I'm not going to change it."

Yoda stood up. Obviously hurt because he couldn't change Ron's mind. "Take a couple days. He's not in any hurry to push it forward anyway."

Ron turned back to his desk. Henderson didn't want to push anything forward that would stir the pot or get his butt in any kind of sling. *Screw Henderson* thought Ron. He needed to have some problems in his life. This was something Ron didn't need to worry about.

The text message was waiting for Ron when he got out of work. He needed to meet Clyde to get a list of appointments for the week. Ron was immediately taken away from his Pentagon office into his other world. It ignited a spark in him that got his blood flowing. He packed up quickly and went to the gym.

After a week of trying to get back into some type of physical groove, Ron thought he was beginning to feel progress. Slowly, his body was figuring out what was happening. He couldn't see any progress, but he sure felt it. He was deliberately not pushing himself too hard, but his body was still sore every night. The punishment was worth it because people were beginning to notice. Even he was beginning to notice the success of his efforts.

CHAPTER 9

5 OCTOBER
"O'REILLY'S TAVERN"
ALEXANDRIA, VIRGINIA

"I can't stay long. Here's a list of this weeks appointments. Can you make all four of those?"

Ron was surprised, but didn't see any difficulties. "Shouldn't be a problem. The only thing I have coming up that I know of is the President's ball next month. I can't believe I have to go to that, but my boss wants everyone in the shop to go. It would cause a bit of trouble if I decided not to go. He's already not too happy with my work performance."

"I see where that's headed. You can't tell the boss that his work is subordinate to anything. You're pushing him into irrelevancy. No boss in the Pentagon, at Microsoft, or coaching in the NBA wants to hear their bullshit."

Ron understood Clyde's point. "Not so much irrelevant, but over funded."

"Oh, the plot thickens! I'm so glad I have you at the Pentagon looking after my tax dollars!"

"Your tax dollars are spent long before they get to me. Just give me the list."

Ron pointed to a name on the paper. "Number one is a guard in the WNBA. I think they're in their off season, so she's just trying to keep in shape for traveling." Ron nodded and hoped she wasn't over six foot tall. "The second one on the list is a CEO for an internet company in Falls Church. She's a beautiful lady, I've seen her. She has never been a client of mine before, so go slow. She was told about us by a friend, so I think she's a first timer. I think she has issues."

"First timer?" Ron asked. Clyde smiled. "Issues, huh?"

Clyde didn't expound. He continued going down the list. "I think you already know Edie?"

"I like, Edie!"

"And this last one is another recommendation. I heard she was a diplomat's wife, but didn't ask any questions. You're basically getting all my new clients to establish your own list."

"Wow! Someday, I might be just like you?"

"Doubt it. But at least you're losing weight!" He smacked Ron on the arm. "Gotta run buddy. Keepin' my list bigger than yours! See ya!"

Ron tucked the paper into his pocket. His list was already larger than he had ever anticipated. With the new list, he was approaching twenty clients. Being with his first few clients had become a routine that he was getting comfortable with. The new ones were a different story. He was never real comfortable getting to know the new ones.

CHAPTER 10

8 OCTOBER
"THE AMERICANA"
ROOM 712
CRYSTAL CITY, VIRGINIA

"**M**y name is Sonja." She was a little older than Ron, maybe forty-five. Yet, very attractive. Ron's first thought was *business professional.* She was about 5 foot seven but the heels she was wearing made her appear taller. The dress fit her perfectly and accentuated her figure. Ron estimated she may have weighed 135. Her hair was shoulder length brown with blonde highlights. She was also extremely nervous. That was something Ron was not used to.

"Hello, Sonja. My name is Steve."

"I've got to tell you, I've never done this before. I don't know how to go about," she struggled to find the right words. "Doing this?"

Ron walked over to the counter by the sink. He didn't want to be too aggressive. "First of all, would you like something to drink?"

"Are you on the clock?"

Ron thought for a moment that she must be a tightwad. Maybe she didn't have a lot of money. He tried to relax her fears. Money had never been a problem so far. This was something else. "I don't have to be. I usually round down anyway. Do you have another appointment?" Then he had a new fear. What if she was a cop? Clyde didn't know anything about her. Was she trying to set him up?

Ron would have to be extra slow with her. "Let's just have a glass of wine and get to know each other for a few minutes."

Sonja seemed more than willing to do that. He took two glasses of wine and sat on the edge of the bed. "So what do you do?"

She took a glass of wine and sat on the bed, but not close to Ron. Sonja was hesitant to talk at first. It was clear to Ron that she didn't think she wanted to be too truthful with the man sitting on the bed. "Look, you don't have to tell me anything if you don't want to."

That cleared the air a little. "I'm a CEO of an international trading firm."

"A CEO?"

"Yes."

Ron waited for more, but she was quiet. He took a small drink and sat back against the head board. The silence was awkward. "So, you want to watch some TV or something?"

She turned around quickly and looked at him. "Is that what I pay you for?" Ron smiled broadly. Was she pressuring him to cut a deal or was she just inexperienced and nervous. It was time to check if she was for real. "What do you want to pay me for?"

She started to get angry and stood up. "I thought I was paying you for sex!" That was good enough for Ron. No cop would have thought of that line.

She headed for the table to grab her bag. He got off the bed and walked towards her. "Sonja." She stopped. There was a tear in her eye. He didn't move any closer, less he scared her off. "I'm not on the clock."

She looked up at the ceiling and wiped her cheek. "I don't know what I'm doing?"

Ron stepped towards her. "You don't have to do anything. I mean, we don't have to do anything."

She turned to him and smiled. "I don't know what to do. How does this work?"

Ron stepped two steps back to give her space. "First, you decide what you're doing." She laughed. "Then, if you are so inclined. We make love like there is no tomorrow." She laughed again. Ron turned toward the bed, set his glass of wine on the nightstand, jumped onto the bed and grabbed the remote. "Or we could just watch TV! I love ESPN at midnight!"

She hesitated for a moment. Finally, the reality of her opportunity set in. This wasn't a marriage. This wasn't even a relationship. It was just sex.

She grabbed her glass of wine and walked to the far side of the bed. She set the wine down on the nightstand and slowly sat down on the bed. Ron turned on the TV. She lay back against the headboard and grabbed her glass of wine. She took a big drink. "I haven't had sex in almost ten years." She instantly took another quick drink.

Ron, initially stunned by the comment, decided some levity was needed. She was wound way too tight. "You wanna to watch Comedy Central? I love that channel!"

She elbowed him. Ron played like he was hurt. "I'm trying to be serious here." He couldn't contain his smile. "I haven't had sex in . . ." he looked at his watch, "about ten hours."

"And that is supposed to impress me? Or disgust me?"

Ron could she that she was taking him too seriously. "I'm being facetious. I don't keep track of that. I'm trying to get you to relax." He took a drink. "I was serious about watching Comedy Central though. I think the 'Daily Show' is on. That's where I get my news from."

She took a long drink. "I'm more inclined to watch some porn!"

Ron nearly spit wine out his nose. "We can do that, too!" She smiled back at him. He decided to push her again. "But if it's working for you, I am on the clock!" She elbowed him again. "Hey, I'm probably cheaper than the darn porn channel."

She put the wine down. "If you're on the clock, you need to put that remote down." Maybe it was the wine, but she finally had enough courage to make a move. She leaned over and kissed him.

"One quick question?" Ron really wanted an answer. "Why has it been so long?"

She exhaled heavily and looked up at him. "I've been working."

"I understand that." She was quit for a moment. Ron broke the silence, "I'm guessing you're not married either."

"Nope. Never was."

Ron nodded. "I was, but not anymore."

"Sorry."

"I was kind of married to my job before, too. So, no need to be. I messed it up."

She tilted her head and asked, "So are you good at this?"

Ron smiled. "I'm not really sure." For some reason he stood up on the bed. He started jumping up and down on the bed and yelled, "But I love work!"

She was laughing as she nearly bounced off the bed. She got off the bed, stepped away and took a running jump and joined him jumping up and down on the bed like two children. She yelled, "If we break it, you're paying for it!" She was a tightwad.

Ron stopped jumping. "If we're going to break it, let's break it in another way."

That was all it took. The line relaxed her and her shell was broken. She moved to him and kissed him hard. She slid slowly down his body, kissing him as she went down his body. Somehow, her clothes came off as she did.

Four hours later, Ron silently left the hotel room. He need not have worried about Sonja's frugality. She gave the appearance of being too concerned about money, but that wasn't on her mind when she was through. She had given him a cashier's check for $1,000.

12 OCTOBER
C RING – 117
PENTAGON, WASHINGTON DC

Ron knew he was up shit creek. Henderson sat behind the desk, tapping a pen on his desktop calendar. The Colonel wasn't screaming, even though he was pissed off. "How can you write this? Do you not understand my directions? What are you thinking about?"

Ron knew exactly what he was thinking. "Sir, I'm thinking about the 19 and 20 year old kids we have in Iraq! I'm thinking about the troops forward . . ."

Henderson cut him off. "You sure as hell aren't thinking about your job here!" He threw the pen down. "Information Operations is just a way that other Directorates are going to try to use and absorb Public Affairs. Between Info Ops and this new Strategic Communications concept, PA is going to be shut down for Christ Sake!"

"Sir, that isn't going to happen and you know that. You're over reacting."

"How can you be so certain, Major?"

Ron bit his lip. "Because Info Ops is about synchronizing operations. It's about coordinating plans, and breaking down walls between groups and organizations. As Officers, we should be happy with that! It's going to help us win on the battlefield."

"We're not on the battlefield, Major Benson. We're in the Pentagon. You need to remember that!"

"I can't forget it, Sir. I remember it every stinking day."

"Let me tell you the benefits of being a staff officer at the Pentagon, Major."

Ron shook his head. "This should be a pretty short conversation."

"I don't like your tone, Major."

Ron still maintained his military bearing. This was not the time to fall on his sword. "Roger, Sir." As was becoming a bad habit, he chewed on his lip again. "I'm sorry." He didn't need to have any problems at work.

"As an officer in the Pentagon, you receive the best opportunity to lead the military into the next decade. This assignment will prepare you for leading troops on the modern battlefield." Ron nearly gagged. "And most of all, it gives you the connections you need to further your career." Ron was stunned. This was coming from the guy that had informed him he was passed over for promotion. That was enough for him.

If Henderson was willing to let him off with an ass chewing, so be it. It was time to leave. "Is that all, Sir?"

"Is that all?" Henderson scoffed. "The Inspection team will be here next week, and they want to talk to everyone on staff. I don't think I want you around here Major. I tried to get you sent TDY somewhere. But I can't get orders cut on you that soon. You wouldn't care to take leave would you?"

Waste good leave time to avoid be truthful. It didn't take Ron long to figure that one out. "They'd just interview me when I got back, Sir."

Henderson was silent for a moment. "I suppose you're right. When the IG team comes through next week, fell free meet with them." He stood up from his chair and leaned over his desk toward Ron, his eyes narrowing for emphasis. "You don't say anything, *anything* that will make me or this department look less than stellar. Do you understand me?"

"No doubts, sir." Ron stood up and saluted. He turned and left without waiting for Henderson to bring his hand up.

He quickly returned to his cubicle and sat down. Inside he was burning with anger. He should have told Henderson to pound sand. He should have told him exactly what he was thinking. He sat back in his chair and looked up at the ceiling. His life sucked.

That was when reality set in. Henderson was a dick, but he was still his boss. Everything Ron wanted to do would have been wrong given the situation. The officer that he was did exactly what he was supposed to do. Take the verbal shots, the threats, and the order to keep his mouth shut. What difference could he make? He looked at his watch. It was 4:20. He turned off his computer and grabbed his gym bag. He could get in at least a two-hour work out before he had to meet Clyde.

Tony was silent as he watched Ron. His cubicle mate was moving at the speed of pissed-off. With a quick wave of his hand and a, "See ya!" Ron was out of the office.

"Bye," was the only word he got in, as Ron was gone. *That was so unlike Ron.* Something was going on, but Tony didn't put it together. It was close enough to five for him to start getting his own bag together for the ride home.

For the first time since he had been at the Pentagon, he was one of the first people at the PAO to leave the office. He didn't know what made him feel guiltier. Leaving work early or the fact that according to the Uniform Code of Military Justice, he was a criminal.

CHAPTER 11

13 OCTOBER
"MCKINSTRY'S STEAKHOUSE"
CRYSTAL CITY, VIRGINIA

Ron could hardly believe his eyes. He could tell it was Clyde, but he was different. Very different. To be honest, he looked great! "Damn, what has gotten into you?"

"I just thought I needed a little change." His hair was blonde, combed back neatly with a small bit of styling gel in it. He was wearing a three-piece suit with pin stripes, a cream-colored shirt and gold jewelry.

Ron was in awe. "No, man. I'm serious. What's up?"

Clyde laughed. "I have a hot number tonight and she's uptown, if you know what I mean?"

"It doesn't take much to figure out what you mean." Ron shook his head and whistled. "You look, fantastic!"

"I have to! You keep getting all these good reports from the clients and you're making me take my game to a new level."

Ron pulled out an envelope and put it on the table. "Speaking of a new level." Clyde took the envelope. "I think there's $775 in there."

"At ten percent," Clyde did some mental math, "Holy crap! You made more than I did last week!"

Ron smiled. "I didn't work that hard. I got a couple good tips."

"Good tips, eh? Your regulars are starting to get jealous."

"I haven't missed an appointment yet. Regular or new."

Clyde shook his head. "That's why I'm getting the new digs." He was smiling when he said it, but it was time to get serious.

Clyde got down to business. "I'm dressed up because I have a special client tonight. She is a diplomat's wife. We're meeting at one of the nicer, secluded places for a private dinner. I don't think the client would appreciate the normal 'Clyde'. Tonight, my name is, I. M. Goode."

Ron couldn't contain his smile. "She must be special, for you to get dressed up like that." The fact that his client was married went by Ron like water off a duck's ass. Things like that didn't seem to matter to him anymore. He had slept with a dozen married women.

"That's just the half of it." He pulled an envelope from his coat pocket and slid it to Ron. "You were recommended to this individual by one of your clients."

Ron took the envelope and opened it up. He read for a minute. "It doesn't say who? It just tells me to meet her on Saturday night."

"Anything else in the envelope?"

Ron turned it over. "No. Nothing."

"All I can say is what I was told. This is a very special client. She was told you were the one she needed because of your" Clyde made quote marks in the air, "'attitude' was the word she used. Now I've heard it called a lot of things, but an attitude? Come on!" Ron laughed.

"I was told she would pay very well and you may have multiple 'engagements' if the confidentiality remains just that."

"I don't see a problem with that. We're not exactly advertising in the Post now are we?"

Clyde grew quiet. "We're starting to work a different group of clients, Ron. Upscale clients. Clients with money. Clients with clout."

"Clients that are horny!"

Clyde smiled. "Clients that need us." He held up his glass and toasted Ron. "To my partner. I wouldn't be getting to this level without you."

"It's mutual, Clyde." He took a drink and put the glass down.

"How's your real work going? You know, the Major part of your life?"

Ron frowned. "My boss is a jerk, the work is monotonous, I'm now the guy nobody wants to sit next to and I get the feeling the chain of cammand would just as soon fire me as deal with any more of my attitude. I wrote a paper that explained what I thought Public Affairs needed to do to help the war, and my boss wouldn't even send it up the chain of command. So I have achieved the apex of a very mediocre career."

"Sounds perfect."

"Life in the Pentagon." He exhaled loudly. "I miss being an officer."

"You're still an officer."

"No. I'm a computer nerd in a uniform. I've been verbally accosted, psychologically abused, and mentally castrated. I couldn't say shit if my mouth was full of it." Clyde laughed aloud. "You go ahead and laugh. My boss basically told me I needed to say what he wants me to say or my ass was gone."

Clyde nodded. "What does that mean? What can he do to you?"

Ron shook his head. "In all honesty, absolutely nothing. I'm not gonna get promoted." Ron got quiet. He looked at Clyde and smiled. "There really isn't much he can do to me. Without the threat of not promoting me, all he can do is keep me hidden and move me down the road."

"Nobody says you have to be smart to full bird Colonel." Clyde let that sink in, then changed the subject.

"What about your retirement? You only need 18 months and you have your twenty."

Ron thought for a minute. "The only reason I want my retirement is so I can pay my child support for Ally. Sandy deserves it too for putting up with all the crap the Army, as well as me, put her through. But the military doesn't owe me a dime. Just like I don't owe the military anymore either."

"Sounds like you got no worries little brother."

Ron still had worries. "The only problem I have is if they find out I'm doing this." He raised his glass for another drink of wine. "If they find out what I do with my spare time, I'm hosed." He took a long drink. "Long tour at Leavenworth, bro!"

"I'd come visit."

"You may be in the cell next to me if your clients have any say in it."

"Oh, no, little brother. They love me." He wiped away some wine on his lip and said, "They love me long time!"

Ron laughed loudly. He may have been dressed up, but he was still vulgar. That was what made Clyde so unique. He held up his glass for another toast. "To us, man-ho's!"

"Man-ho's!"

That was the last drink for the night. They both had appointments and Clyde couldn't afford to be drunk for his special client. Ron was more subdued by the wine than normal and took his time with his regular customer. He stayed longer than he wanted to, but the last hour was on him. He cursed himself as he left for staying so late. He had not intended to stay out past one, but he felt better being there than alone in his apartment.

On the way home he thought about the note Clyde had given him. It was actually quite mysterious. There was something exciting, yet dangerous about this new adventure. He didn't know whether to look forward to it or call it off.

CHAPTER 12

18 OCTOBER

THE ROYAL CROWN PLAZA HOTEL

WASHINGTON, DC

"I believe you might have a message for me. My name is Steve Jones."

"One moment please, sir." The concierge looked around his counter. He found what he was looking for. "Here we go, Sir."

He handed Ron a legal envelope with instructions inside. He was told to go up to the twenty-eighth floor and wait for two men. He would be searched then told what to do.

"How do I get to the 28th floor?"

The concierge pointed and said, "Right over there, Mr. Jones. Last elevator on your right."

Ron nodded and headed toward the elevator. The ride was actually very fast. Once he reached the floor, he stepped out of the elevator and looked at himself in the mirror. The collar felt tight around his neck and he wasn't certain that the tie matched. It would have to do. He nodded approval to himself and hoped that his client would approve as well. If not, it was a heck of good effort.

There were no men waiting for him so he stood next to the elevators in silence. It was the penthouse floor and he noticed there were only four apartments. He saw a large couch in the foyer, but decided not to sit. For some reason, he was nervous.

He wasn't there long. The two men that came down the hallway were big. Both were at least six foot two and two hundred pounds. One gentleman in the front was black and he looked familiar to Ron. The second gentleman was Caucasian and probably had been a Marine. The crew cut was a dead giveaway. Ron wasn't sure, but at first glance, they looked like secret service agents.

"Mr. Jones?" asked the large black man in the front.

"Yes." Ron stood still with his hands to his side.

"We need to search you to be sure you don't have any weapons or recording devices." Ron understood the search for weapons, but the recording devices comment caught him off guard. He had to think for a moment if he had any. His cell phone was great at any location, but didn't record anything, not even pictures. "Could you loosen your belt and unbutton your shirt, please?" asked the second agent.

The first agent watch carefully as the second agent went over Ron's body with a metal detection wand. He looked at the black man and a light of recognition came on. "Is your name Gault?"

The agent smiled, "Yes, Sir."

"Rollie Gault?"

"Yes, sir." Gault was actually surprised he had been recognized.

"Damn! Fancy meeting you here?" Ron smiled as he remembered the younger man. The second man made Ron lift his arms as the search continued. "What's it been six years now? I was rooting for you guys like crazy! I thought you had Kansas that night." Roland Gault was the starting point guard for Georgetown University in their last Final Four appearance. "I picked you guys to go all the way. No offense, but I lost my office pool because of that game."

"We were pretty good. It just happened that Kansas had Donovan Taylor and he was a lot better."

Donovan Taylor was playing power forward for the Boston Celtics. "Oh, yeah! He was something. How 'bout you? Did you get a tryout?"

"No. Tough cut to make it in the NBA."

"So I'm guessing you're a secret service agent now?"

"Yes, Sir." Roland Gault got back to business. "And you are?" He let the question hang in the air.

Ron was smiling and almost said "Major". But he could see Mr. Gault was a professional. "I'm a going wherever you tell me to go, Agent Gault."

Gault smiled. Not the answer he wanted, but that would do for the moment. "Thank you, Sir. That will work out fine."

"He's clear," declared agent number two.

"And I'm guessing you were a Marine at one time?"

The Marine was much more formal. Perhaps he knew what Ron was doing there and he was none to pleased to be compromising his integrity with this 'detail'. "I was with Marine One and got an opportunity to transfer to the secret service. Seemed like fine honorable work." There was a touch of disgust in his voice.

"And your name Sergeant?"

"My name is none of your damn business." He stepped back and extended his hand down the hallway. "Right this way, Mr. Jones."

"Semper Fi, Mr. Damn Business." He looked at Gault. "This one's the joker, right?"

"He's not always like that." Roland Gault turned to Ron. "Follow me please, Sir."

Ron followed Roland Gault closely. He couldn't help but feel like a human sandwich with the Marine looking agent very close behind him as they walked down the hallway. They came to the end and Gault unlocked the door and entered first. He called out, "Ma'am, Mr. Steve Jones is here to see you."

From somewhere across the penthouse he heard, "Fantastic, Roland. Have him wait in the living room. I'll be out in about five minutes."

"Mr. Jones. Be advised that if we hear anything . . . unnatural. Anything strange or just plain become too curious during your activities, we will be in here within two seconds."

The Marine snarled, "We'd hate to have to come in here and kill you."

"I would hate that too, Gunny."

Gault pushed the former Marine agent out the door. "Come on, Agent Klein. We will be right outside the door, Mr. Jones." With that, they were gone and he was alone on a landing that faced a huge living room.

Ron walked slowly into the large living room. It was hard to believe they were on the 28th floor. The ceiling was at least twelve feet high. There were at least three six foot tall paintings on the walls. He didn't recognize the artists, but was certain they were originals. He wondered how much an apartment like this cost? Was this apartment a nightly deal or did his new client own it?

He walked over to the bar. Considering the situation, he thought he should be polite. He raised his voice a bit to ask, "Would it be all right if I had a drink?"

"Not a problem," came a voice from somewhere down the hallway.

"Would you care for one?" He shouted.

His new client walked around the corner. "Why thank you. I would like a Gin and Tonic, please." Her voice became clearer as she walked into the room.

Ron recognized her immediately. His special guest was the junior Senator from the state of Maryland, Denise Mitchell-Samuelson. She was wearing a man's buttoned down-shirt with pink flowered pajama bottoms and nothing else. She wore no make up, her shoulder length chestnut hair was down and she probably already had a drink. Maybe two.

Senator Mitchell-Samuelson walked right up to Ron and stuck out her hand. "Do you know who I am?"

Ron smiled broadly as he took her hand. "Yup! Do you know who I am?"

She laughed. "I'm Denise. And no, I don't know you. But, I've heard a lot about you. How are you, Steve?"

Ron was struck silent. He shook his head and smiled, "I'm . . . I'm kind of stunned right now."

"I was hoping you wouldn't recognize me."

"We have C-SPAN on at my . . . ," he cut himself off. "I've seen you on TV."

"I was afraid of that. Is that going to be a problem for you? You know? Down there?" She pointed at his groin.

Again, Ron was shocked. "I . . . have . . . no idea." He took a big drink of a fresh Seven and Seven. "I'll let you know in a while."

She took her drink and turned to walk towards the couch. "I figured I would dress down tonight. I'm sorry you went to the trouble of dressing up."

Ron pulled on the front of his blazer. "Dressed up? Me? This is nothing? I always dress like this."

"Right. And I'm not a Senator!" She took small sip of her drink and sat down on the couch with her legs crossed Indian style.

Ron walked over to the couch and sat on the far end away from her.

"You can sit closer. I won't bite."

Ron nodded. "Maybe in a minute." He was trying to gather his bearings and get a clearer picture of what was going on. "You're on the Senate Armed Services Committee aren't you?"

"Yup."

"You're a Democrat, right?"

Ron could tell be the look on her face the comment didn't sit well with her. "I didn't know you were going to be so politically in tune. Do you have a problem with that?"

Ron seriously considered the question before answering. "Not really. I don't have anything against Democrats." He bit his tongue. In fact, sometimes he did have something against Democrats. Like unbridled elitism and unquestioned hypocrisy, those in congress demonstrate. Hell that went for the Republicans as well so his answer was true. For just a moment he felt like he was back in his office in the Pentagon starting to watch what he was saying.

Then it hit him. He wasn't in the Pentagon. He was at his new job. A job that was more about the truth than his old one. "Well, not that you're a Democrat exactly, but that you vote that way." He struck an open nerve with her.

She stood up. "Hold on! I'm not here to talk politics. You're not here to talk politics either. I'm putting that away for a few hours and if you can't, then you can go right now!"

Ron nearly crossed a line with his new client. Denise titled her head as she asked, "Are my political views so bad to you? Is the fact I didn't get dressed up for you the problem?" She gave him an up and down glance and added. "Or maybe you can't get it up now?"

She was rather crude, in a playful, bitchy sort of way thought Ron. He needed to get control of the situation. He immediately smiled. "That's not a problem, Senator." He got up off the couch and finished his drink. He turned and walked back toward the bar for another. "And by the way, your views are not all bad. I can put away the fact that your political views on most issues are wrong if you can." He turned and smiled at her. He started to wonder if he would see here legendary temper.

"Oh this is sweet! You don't think you can have sex with me because of my liberal stand on issues. What kind of escort are you, Steve?" She raised her eyebrows as she said, "I think I got a bad piece of advice from Sonja?"

"Sonja? Sonja recommended me to you?"

"She happens to be a very close friend of mine."

"And a big contributor to your campaign fund, I'm sure." He bit his tongue again. That was a comment he should not have brought up. This was not how the meeting was supposed to go. He turned away from her and took a deep breath and waited for her to explode.

To Ron's surprise she didn't come close to being anger. After a few seconds she put her drink on the coffee table and said calmly, "If you must know, yes, she is." She walked toward the bar to join him.

"My views are the views of my constituents. My vote is for the people that put me in office. Sonja helped put me in office." She stopped behind him and held out her glass. Ron turned and grabbed the glass. "She happens to think you are very good at what you do."

He smiled at the compliment. It made him relax. "I guess I am. Sonja makes it easy. She's a whole lot of fun when you get through that professional exterior."

"I'm fun when you get through my exterior." She moved closer to him.

"Well let me say with all sincerity. I'm sorry." Ron let the comment sink in. "I will not bring my political views into any further conversations we have. I apologize for the earlier distracting dialogue."

She looked him up and down. "Apology accepted, Steve. Are you ready for some fun?"

Ron nodded and smiled shyly. He was not ready to touch her yet. Slowly, he walked by her and glanced down at her ass as he walked by. He held the glance long enough to make sure she caught him looking at her.

Ron directly behind her and stood only inches away. She always appeared to be a little heavy on television. As he looked at her, he decided that for a woman in her mid 40's, she wasn't heavy at all. She was actually very attractive for her age. He wanted to ask her how she kept in shape, but thought that was a line to be used by amateurs attempting a pick up in a smoke filled bar. He wasn't an amateur and it was time to prove it.

First, Ron tried to put away the fact that one of the most powerful women in the world was mere inches away. He put away the fact that he vehemently disagreed with her politics, her social opines and her ruthless ambition. He wanted to ask her about her husband but knew that was irrelevant at this point. Then he put away his drink.

Ron placed both his hands lightly on her shoulders. "I didn't think I was here for fun. I'm also certain that I'm only going to get through the exterior you want me to." He wanted to smile at the innuendo, but that would have been as crude as her and a definite mood breaker. He rubbed her neck and noticed her head tilted to the side at his deep touch.

Ron moved closer to her and she stepped backward to meet his advance. Ron moved his left hand around her waist and pulled her tightly against him. He was not as surprised as she was when his erection pressed against her.

That was all it took for her. She was no longer a Senator. She was a woman. A very lonely woman that didn't need Ron. She needed anyone. Anyone that would hold her and touch her the way a woman needed to be held. Anyone that would allow her to be a woman again.

Ron pushed her away slowly and turned her face to look into her eyes. A thought occurred to him and he smiled broadly. Denise looked up at him and smiled. "What are you thinking?"

Ron slowly bent over, then, unexpectedly picked her up over his shoulder. Again, she surprised him. Senator Mitchell-Samuelson burst out laughing. He grabbed a bottle of something off the bar and carried her to the bedroom. He slammed the bedroom door and waited for a few seconds to see if either Agent Gault or Klein would come into the Penthouse.

When neither agent appeared, Ron understood they were really going to be alone. Just a man and a woman. From her initial actions in the bedroom, it was obvious that it had been quite some time since Denise Mitchell-Samuelson had been 'just a woman'.

It was not a smooth intercourse at first. They tried a couple of positions, but none seemed comfortable. There was initial awkwardness that he had not experienced with his previous clients. After a few minutes of exploration, he took charge of the exchange and put her on all fours facing the headboard. That was when he started experiencing unwanted thoughts.

While trying to maintain his composure and stamina, Ron found himself thinking about tangents. Thoughts of who she was, how she had always made a habit of voting against the military. Numerous votes against funding and resources kept popping into his mind. Times he had seen her give speeches against the military. He found his strokes became stronger, as if anger had taken control of him. He didn't like what he was thinking.

The thought that he was screwing a member of congress that had indirectly screwed him throughout his career was not lost on him. He literally shook his head to clear his mind. The negative stream of thought had made him lose focus and Denise briefly looked over her shoulder at him.

There were no words spoken. Nothing needed to be said. Ron put a hand on the back of her head and softly pushed her towards the pillow. He heard her exhale and started to meet her rhythm. Ron was glad he had been with other women before her, because she testing him. He couldn't help but think this was for every soldier in Baghdad and Kabul and every other hell hole the government had sent them into.

Finally she lost her control. The only way Ron could initially tell she climaxed was by the way her hand smacked the headboard. She made no sound, gave no warning, just a hard slap against the wood, followed by a small groan as she raised her head to the ceiling. He could feel her body tighten, and then relax as the waves rolled back into the depths from where they came. He was still and let her enjoy the feeling of passion as it swept her away.

To his surprise she wasn't done. She forced him onto his back and climbed on top of him. The exchanges continued, like waves on a shore. No longer a contest, but a love making session that became better for each as unspoken barriers were knocked down.

Somewhere around midnight, Ron forgot who the client was. He forgot her power and her stature. More importantly, he forgot about his earlier negative thoughts, of his revenge against congress for all the miseries of the Defense Department. That was not the fault of the woman that was with him.

He found himself getting lost in the moment. As she made love to him, he realized that he was supposed to be the professional. As the night went on, she was becoming a teacher. It may have been a long time for her, but she had remembered lessons she must have learned in her past and put them to use. She made love with abandon. After the initial rust was gone, she proved to be more than experienced. In fact, she was damn good. He found himself wondering how much he would have paid her only a month ago.

Around two in the morning, Ron quietly arose from the bed, grabbed the bottle and walked to the window. He took a small drink and to his dismay, he found it was bourbon. It managed to wake him up a bit when the warmth of the whiskey made its way down to his empty stomach. He opened the curtain and looked out into the Washington night.

A light snowfall had started. This time of year in the District was always the best. The covering of snow seemed to take away all the cities filth. It was a lovely town when the snow cleansed it.

Denise sat up in bed. "What are you doing?"

"I'm sorry, I thought you were asleep."

She climbed out of bed and put a silk robe on. She walked behind him and put her arms tightly around him. "Looks awful cold out there."

"It's a cold town," said Ron barely above a whisper.

Denise agreed and said with a little smile, "Don't I know." She knew way too much about how cold the city really was. She had been a Senator for over ten years. Ron couldn't remember how much more time she had spent in politics before she came to DC. Long enough to be well versed in all the political games the Senate mustered. He thought of something else. "I was going to ask you about your husband."

Her grip around his waist loosened. "Don't."

"Is it working out?"

"For now. He's a Billionaire. I'm not sure if he married me for status or if I married him for money. It was so long ago, I don't remember anymore."

"I don't want you to worry about whether or not anyone will find out about this. I have my own reasons for wanting to keep this just between us."

"Sonja said there was something 'mysterious' about you. She said you were very private."

Ron looked outside. "I'm a product of this city. I need to keep what I do a secret, too."

She nodded approvingly. "That works for both of us. So are you still on the clock?"

Ron turned to face her. "I don't have anywhere to be for awhile. Did you have something in mind?"

She grabbed his hand and starting walking toward the bed. "Something will pop up." When they got to the bed, she turned and grabbed his groin. "See."

At 5:35 Ron woke up alone. He rubbed the sleep from his eyes and quickly got his bearings. The Senator was already gone. She had left sometime in the early morning and Ron had not even noticed she was gone. He decided to take advantage of the penthouse surroundings, ordered room service and showered.

When he went to the kitchen, he found an envelope with ten new $100 bills. For a second he was mad at himself for all the negative thoughts he had during the first hour with Denise. For a time, he had forgotten he was doing a job. His emotions were scrambled because he felt like a jerk after the first sexual encounter with her.

She wasn't a bad person. Voting is what she did. If that meant screwing over some fellow Americans, she merely represented people that wanted them screwed over. He shook his head at his own absurd rationalization. Screw her job. Screw her politics. He genuinely liked the woman that he had brought out last night, not the politician she was everyday.

The money was just a nice bonus that capped an incredible evening. He vowed that if he ever was with her again, he would not bring up politics. She was a client and he needed to separate her political status from his work. As he left the penthouse, he hoped he would have the opportunity to test his newfound resolve again.

CHAPTER 13

20 OCTOBER

2E800

PENTAGON

WASHINGTON DC

"Hi Sandy! I'm surprised that you called me at work." Ron suddenly became afraid. "Is everything all right with Ally?"

Sandy Benson quickly calmed his fears. "She's fine, Ron. I just called to say thank you so much for, not just being on time, but having the extra money with the check this month."

"No problem, Sandy. If our daughter wants to go to Princeton, we better start putting some extra money aside, ya know?"

"Yes, Ron, I know. I'm hoping that David will help with her education." David was Sandy's steady beau for the last year.

Ron didn't like David Weston. Ron found himself pissed that the guy had not asked Sandy to marry him. *But why should he when Ron's alimony kept coming.* He was angry at himself for thinking that way, but he couldn't help it. He put his anger aside. "I'm hoping he will, too." He waited to see if Sandy had anything else to say, but she was quiet. "We'll get her through somehow, okay?"

"Yes, we will, Ron." She was looking for something else to say but nothing came to mind. "I better let you get back to work. I'll talk to you again soon, Ron. Bye." She quickly hung up the phone.

Ron knew better than to expect anything more than that. He still loved her, as strange as it may have seemed. He knew she would never take him back. He had taken her trust and traded it for a career in the military. A career that had left him alone. He may have broken Sandy's heart, but the Army had ripped his out. It wasn't a fair trade. Nevertheless, he had been told back in Officer Candidate School the Army wasn't fair.

It was that phone call that made Ron come to a realization. The military life was meant for single men. Military men were better off without wives. The military was made for men without hearts. It was made for men that don't know or want love. It wasn't made for Ron Benson anymore. He began to wonder if he was rediscovering passion. Maybe even love? How much longer could he stay?

25 OCTOBER
2E800
PENTAGON
WASHINGTON DC

"Ladies and Gentlemen, this is Major General Walter Kusman. He is the leader of the investigating team. General Kusman," said Colonel Henderson. "The floor is yours."

Major General Walt Kusman was a Marine with combat experience from the first Gulf War. He was a no-nonsense kind of man and he didn't take fools lightly. Whenever he mentioned the term 'Public Affairs', it appeared that he was in pain. Just by mentioning PA, the General was putting words in his mouth that he had spent a career avoiding. It was obvious to Ron that he would rather have been forward deployed in the Green Zone with a battalion of Marines rather than be talking to a bunch of candy-ass Pentagon staffers that probably had no idea what war was.

As expected, his in-brief was short and to the point. "Thank you all for being here. This get together is to discuss how you all conduct Public Affairs, how you think it's working out and how, if possible, we can make it better. I will be meeting with each and every one of you as time permits. Please think about why we're here and be ready to tell us what you think."

Henderson was watching Ron during Kusman's entire introduction. Henderson was positive that if Ron got the chance, he would do or say something to the General that would make him or the Office look bad. If there were any way possible to have the Major out of the area, he would have done it. Ron squirmed slightly in his seat. Kusman may have wanted to know everything that Ron thought, but H2 No wanted Ron to keep his mouth shut.

When Ron got the news that he and Tony Parkman would be interviewed together, he thought he'd be able to slide. Maybe he could keep his mouth shut and avoid Henderson's wrath. Some things are just never meant to be.

Late in the afternoon, the General finally got around to Ron and Tony Parkman. He was short and appeared to be some combination of bored shitless and tired. Kusman didn't beat around the bush. He was a warfighter. Interviewing PA pukes was his mission and he would do the best he could with it.

"What do think of Public Affairs role in the military, Major? Or should I say, Majors?" Kusman came out firing and directed the question to either Major.

Tony Parkman was nervous but jumped on the question immediately. "I'd have to say there is a place for PA in the military, General. We provide a valuable service to the media and the warriors on the front lines."

Ron wasn't sure but he thought he heard Kusman grunt. "And you Major Benson?"

Ron was hesitant as he carefully chose his words. "I agree that PA is valuable. I think we are at our best when we team with other departments and maximize our capabilities." Ron nearly laughed out loud at his own voice. 'Team with other departments' and 'Maximize our capabilities'. He was so full of shit he thought his blue eyes had turned brown. He wondered if Kusman could smell it?

The general looked at Ron and said flatly, "Right". Kusman was thinking, *I came over here to waste my time on this?* Yet he tried to maintain his military bearing.

At first Ron couldn't read his expression. Was it was one of confusion or disgust? The General leaned back in the chair, titled his head and frowned. Ron knew right then that General Kusman had caught the foul stench of bullshit.

The General dismissed Ron's answer and went back to Parkman.

"How do you think PA can improve?"

Parkman was all smiles. "I'd say we're pretty damn good the way we are, General. We're actually special staff to the Joint Chiefs and we're in the loop on all their activities, like their travel and policy directives. We help keep them abreast of the situation in forward areas through coordination with other folks. Like Combat Camera."

Kusman responded again with, "Right." He turned to Ron. "Your thoughts, Major Benson?"

"I'd agree with Major Parkman, sir."

Kusman had heard enough. *Do these guys think I'm a fool?* "Does Major Parkman have his hand up your ass, Major Benson?"

"Sir?"

"Either you really do agree with his statements and you're a puppet and he's making your mouth flap, or you actually have a mind of your own and won't tell me what you really think. Which is it, son?"

Ron was debating internally how to be less fucked than he was. If he said he was a 'puppet' of Tony Parkman's, he might as well just kiss the General's ass. If he told him what he really thought and word got out, Colonel Henderson would most likely fire him or have him sent him to Antarctica. The thought that probably both vengeful acts were coming his way crossed his mind if he said a word weighed heavy.

There was the third option that for Ron. An almost unheard of option. Tell the General the truth. Tell him what he really thought. If he was going to get his ass handed to him by his boss, maybe he should give him both cheeks. At least he'd be able to look at himself in the mirror.

Honesty is the one trait the Army requires that separates the soldier from common man. Ron still had enough soldier left in him to remember this. Ron rubbed the carpeting with his polished shoes, cleared his throat and exhaled loudly. The truth couldn't hurt.

"All right, General. You want to know how we can improve." The General merely smiled and sat back in the chair. Ron was on his way down a path he couldn't return from.

"General, I agree with Major Parkman because I was told to agree with him, sir." At first Tony Parkman suppressed a smile. Then it hit him what Ron was doing. His face flushed red, but he knew better than to speak.

"Colonel Henderson told me I was to stay on board and say nothing derogatory about PA, sir."

"I see."

"Can I ask you something, General?"

"Certainly."

"Did you request to meet with everyone as individuals or did you request to talk to us in pairs or small groups?"

Kusman smiled. "Your office decided to save me time by having me talk to you in pairs. They laid out the schedule as it Major Benson." Ron's silence was betrayed by eyes that told the General he had lots more to say.

Kusman turned to Tony Parkman and said, "You're free to go, Major. I'd like a few minutes alone with Major Benson."

Tony Parkman stood and saluted the General then looked at Ron. His face was red. Henderson had probably ordered him to keep Ron in check. Parkman had failed another mission for H2 No. There would be hell to pay for not overseeing Ron's responses. The question was whose ass would Henderson chew first?

The general put a stick of gum in his mouth and pulled out a pencil to take notes. "Now, Major Benson, would you care to answer my questions again."

Ron smiled and pulled open a notebook. He handed the General his paper that now had a title: "Improving Public Affairs to Serve the Warfighter". It contained sections on Information Operations and how it related to Public Affairs and how PA could be used in the 21st Century. He added a new piece that indicated Strategic Communications support for Public Diplomacy was seriously lacking. Another section indicated how Ron felt Combat Camera was being improperly used. In Ron's mind, a more aggressive approach with pictures, and video showing the atrocities of the terrorists in Iraq and Afghanistan would probably work better than the current program showing schools being built and Iraqi electrical workers standing around in power plants. He handed the ten-page document to General Husman.

"I think you asked how we can improve. The paper pretty much explains it, General. If we are to be of better use to the Department of Defense, we need to be wholly integrated into Strategic level operations, all the way through to tactical ops at the Regional Command Level. We need to be a proactive department with full spectrum media operations from camera's to radios to laptop computers with blog sights. We should be televising on Satellite from all the geographic command locations showing the terrorists for what they really are. Not just showing visual images of the positive things we do. Like the support we did for the Tsunami victims in Indonesia. We should be pushing information from all levels in all directions.

"Right now, the entire military, not just the PA office is in a 20th Century mind set fighting an enemy that is using 21st Century media daily to kick us in the ass. We're resourced to do that, but we don't. This new directorate, called Information Operations has the concept and we need to explore it. They integrate, coordinate and synchronize information that benefits guys forward. We aren't doing that here."

Kusman was looking at the paper and smiling. He looked at Ron and the smile broadened. "Do you have a few more minutes to talk, Major Benson?"

Talk is what Ron did. For forty-five minutes, he explained everything he knew about Info Ops, PA and how it related to the warfighters and strategic communications, or lack of it, at the Pentagon. To his surprise, General Kusman took at least six pages of notes.

When he was done, Ron told the General, "Sir I've just crossed a line that I can't jump back over. My boss will know it was me that gave you this information."

"Major, I don't think your boss will do anything to you. There is nothing wrong with crossing a line like this. As long as you're not going to go over it buck naked!" The General laughed heartily at his own joke.

The comment would have been funny in other circumstances. The irony of the fact that he had crossed a military line by literally being naked quite often in the past few weeks was not lost on Ron. He could only smile.

Ron's perspective of what he viewed as the truth had left had just been written in a Generals' notebook. That little flicker of hope that he could stay in the military had grown dimmer. The only thing that remained was one serious ass-chewin' from Henderson.

27 OCTOBER
OFFICE OF THE DIRECTOR OF PAO
PENTAGON
WASHINGTON DC

To Ron's surprise, Henderson wasn't screaming. His face flushed beet red with anger as he spoke. "I received the verbal out brief from Major General Kusman. Needless to say, you left an impression with him." He stopped pacing around the office, turned and faced Ron. "I don't know how you managed to get alone with him. That imbecile Parkman was told not to leave you alone with him. It was so nice of you to give the General that damn paper." *I ought to have you court martial for insubordination.*

Ron twisted in his seat and cleared his throat. "The paper was going to get out someday, Sir."

"It didn't need to be during this investigation!"

Henderson continued pacing. "Here's the deal, Benson. I'm going Temporary Duty to Leavenworth to discuss this investigation with the Combined Arms College folks. While I'm gone, you are not to talk, walk, move, sleep or shit around General Kusman! Or any of his staff. You read me?"

"Yes, Sir."

"If I hear anything negative about you, you are gone. You are reviewing photos of Medical Operations in Mongolia! You got it?" The thought of watching Army medical Doctors stick needles in crying orphans in Mongolia didn't frighten Ron, but he knew any Medcap mission would suck. Henderson's point was taken. The problem was, Henderson wasn't through.

"You are to report up to Major General McCoy's office immediately. I believe the General would like to speak with you. I'm assuming that will be a one way conversation."

Ron started to say 'that won't be necessary'. It would have been easy to just say, "I'm done". Write a letter of resignation and walk away. He knew that going to see McCoy meant Henderson was tired of dealing with Ron. He thought ahead to what McCoy would do. Then it hit him. It couldn't be much worse than Henderson's little tirade. If he made too much of a big deal about it, Ron could go back to Kusman. It would be a chicken-shit way to deal with the situation, pitting one General against another, but that seemed to happen regularly in the Pentagon. The thought gave Ron a little more confidence.

Screw it. Mongonia ain't so bad. "Roger that, Sir." Ron saluted and did exactly as he was told.

27 OCTOBER
C RING – 427 OFFICE OF THE DIRECTOR OF PUBLIC AFFAIRS
PENTAGON, WASHINGTON DC

Ron Benson wasn't in the habit of being in General's offices. He somewhat enjoyed the anonymity of life in the Pentagon as a mid-ranking officer. That anonymity was gone now. Giving his paper to Kusman had me him stand out. Which in itself wasn't a bad thing. Unless you were standing out for embarrassing your entire chain of command. He might as well have been in a Cessna over Tehran because he was on McCoy's radar with missile's tracking. The only question was whether General McCoy would pull the trigger.

McCoy was actually much calmer than Henderson. He sat back in chair and skimmed Ron's paper. Ron, trying to appear in control, sat at the position of attention directly in front of the General's desk in silence.

Soon enough McCoy put the paper down on his desk and leaned forward. "I guess I'm not surprised to see this, Major Benson. I'm just damned surprised to be getting it from Major General Kusman."

Ron bit his lip and remained silent. McCoy's eyes closed slightly. "You honestly believe that crap you wrote don't you?"

"Yes, sir."

"Let me get this straight. I understand that whole strategic thing. That is so above your pay grade that you have no clue, Major." The general looked over top of a pair of reading glasses as he if he were a teacher scolding a student. Ron was silent and remained as expressionless as he could.

"Proactive Public Affairs. You want to try and teach me something about that?"

Ron titled his head to one side. As if his physical movement could avoid the verbal joist that had just pierced him. "I don't think there is anything I can teach you about proactive PA, sir." It didn't work.

McCoy smelled blood. The Major was going into passive defense. *He's not so full of himself now.* "And aggressive combat camera? In this paper," the General picked it up and threw it back down on his desk. His voice changed and Ron detected a bit of condescendence in his tone. "You think a bunch of pictures is gonna win this war?"

Ron started to become angry. Pictures were winning the war for the terrorists. Pictures on Al-Jazeera, pictures on the internet, some true, most fabricated. Ron's eyes narrowed. "Not by themselves." He started to bite his tongue, but couldn't. "If they were put out there for people to see in a," he searched for words that would not indicate current failed policy. "A more expeditious manner, on a website. Or by media that was more favorable to our activity."

The general could sense Ron was no longer accepting his ass-chewing gracefully. "Are you saying our media are not demonstrating favorable US military actions, Major Benson?"

Ron nearly blurted out "DAMN RIGHT!" He took a deep breath and gauged his response. "I don't believe the media have the US military's best interest in mind, Sir."

"You have proof of that?"

Ron's thought's ranged from '*he's just screwing with me*' to wondering if the General was a complete fool. Again, he backed down before his mouth got him into more trouble. "No, sir. I have no physical proof of that." There was a moment of silence.

"Is that part of the animosity towards the embedded reporters you write about?"

Ron nodded his head in agreement. "Yes, Sir, it is." He started to expound on what he meant, but could feel the General didn't want to hear anymore. In fact, the General was getting pissed.

"Go on. Explain it to me. Enlighten me Major with your omnipotent wisdom."

Ron let the snide remark go and focused on answering the question. It was time to sink or swim. He exhaled and looked the General in the eye as he spoke. "The embeds are employee's of big US companies or predominantly global corporations that are liberal, and predominantly anti-war mouthpieces." Ron broke it down to layman's terms. "The feedback I got was they usually stay in the green zone, hang around the chow hall and make a lot of shit up that they think their producers want to hear." The general was silent, so Ron continued.

"When they do go out, the reporters tend to stay in armored vehicles, becoming burdens to the troops that have to alter procedures to keep them alive. Although the reporters think they are doing our guys favors, their work doesn't get pushed onto the front page or the small screen of nightly news unless it's what producers want to read, see or hear.

"At the end of the day, our guys get them home and they end up sending stories back that are anti-US, anti-Iraqi or neutral on the insurgents. The reporters go home without having to explain to the troops or the Iraqi population how come no good news comes out." Ron caught his breath, but he couldn't stop. "The reporters sell us out. It's bullshit, sir."

The General exploded. "I'll tell you what's bullshit, Major!" He stood up and walked toward his window. "You're over here in the Pentagon, for Christ's sake! How the hell do you know what's going on over there? Do you have any idea?"

Ron stuck to his guns. "Yes, sir, I do! I've got friends in Baghdad and Kabul, sir. I get emails every day. Emails that ought to be pushed out to the media. Emails with pictures that counter the anti-administration propaganda that's pushed on the public everyday! Yet we sit here and listen to this crap, taking major media bullshit like a donkey in a hailstorm! Too stupid and too stubborn to do anything but stand there!" Ron got up out of the chair. He was certain what little chance left he had to stay in the military for a 20-year retirement was over. *Hell, the retirement check isn't worth this bullshit. Might as well go out fighting.*

Ron had nothing to lose so he turned loose his thoughts with both barrels. "You want to know if I have any idea. You're damn right I do, General! It's to get out from behind the computers, get the facts and push them out to the public. America needs to know what we're doing! They don't need to get their news from Al-Jazeera and US news outlets that want to see us lose this war because they have a hard on for the President!"

Ron was on a roll. He had newfound courage and drilled his message home. "We're fighting this thing with half the money we need, half the manpower, half the congress and only half the country!" He let the words fly across the room then added, "So as far as I can tell, we're fighting it half-assed, General. And even with all that, our guys are kicking the shit out of the bad guys, which is what their job is!"

General McCoy turned and yelled, "Don't lecture me, boy!"

Ron came to the position of attention, looked at the general and said calmly, "Then do something about it, General." McCoy's face was red with rage and he was visibly shaking. Ron's voice was lower when he said, "You're in a position to." Ron shook his head. "A Major in the Pentagon makes slides and writes papers, sir." Ron stopped talking and looked straight ahead. "Slides and papers don't win wars, sir."

McCoy walked directly in front of Ron. He exhaled loudly as he got control of himself. "I'll say one thing for you Benson. You got balls." He stepped back and relaxed even more. "I guess what really pisses me off is that I didn't get this paper from you. Or my staff." Ron knew Henderson would get his ass ripped for not forwarding the paper and suppressed a smile.

McCoy exhaled and shook his head. "I'll take some of your thoughts into consideration, Major."

"Thank you, Sir."

"You are by no means excused from your action." The general moved behind his desk and sat down. "I will have to think about what your punishment will be. Until I do," he leaned back in the chair. "Get the hell out of my office. I don't want to see you in here again."

"Roger, Sir." Ron saluted and did a quick about face. He was screwed, but not as bad as he could have been.

McCoy turned to look out his large window that overlooked the Potomac. He wanted so badly to bring Benson up on some kind of charges. Any kind of punishment that would put the major in his place. But how he could bring charges against a junior officer for saying the same things he felt. Besides, he had given Kusman his word there would be no repercussions.

That wasn't what was really bothering him. Deep down McCoy knew exactly why he had reacted so tersely with Ron Benson. McCoy knew they should be fighting a proactive war. He knew they were fighting with one hand behind their back. He was just as pissed as Benson at the American media. McCoy knew Bensons' ideas were on the money because he had nearly identical ideas months earlier.

The difference was, the Major's information was not pushed up to him by his staff. The previous August, McCoy had sent nearly the same concepts up his chain of command. He had personally handed the folder to the Assistant Secretary of efense for Public Affairs. For his effort, McCoy had received the same type of treatment that he had given Benson, only it came from a civilian. It was a meeting that he would never forget.

McCoy had explained proactive Public Affairs and the need to be aggressive in getting the truth out to the world before the enemy did. The enemy he thought to himself. Nothing more than kids with laptops, video cameras and an idea. He had the same thing and a multi-million dollar budget. Throw in the Assistant Secretary of Defense for Public Affairs, Washington DC politics and all the bureaucracy the Pentagon could muster and you get an information cluster fuck. It pissed him off that some jihadi prick with a video camera was taking pictures of Improvised Explosive Devices going off on American convoys and he couldn't even show a drop of blood that Al Qaida was shedding. The DOD was winning the war, but losing the audience. Without the audience to mold popular opinion, it didn't matter if they captured every insurgent in Iraq. The appearance was the US military was losing.

As for punishment for Major Ron Benson, his decision was not to not make one. Benson was an upstart and one thing about them, they were usually screw ups. It was only a matter of time until a screw up does something stupid. *This guy will do something stupid. I won't have to wait too long. I need to have someone catch him.* McCoy decided the Major might need someone to oversee his activity, inside the building as well as outside. For a moment, he thought about not making the call. Letting the whole thing fade away like so much other bullshit in the building. The Major shouldn't be punished for doing the same thing he would have done if he could. He hadn't had the balls to retire back then and wouldn't ask Benson for his resignation now either. Unless they started taking away his budget for other new start up programs, this event didn't need to go any further.

McCoy concluded that Benson was so damn smart, he would figure out before too long what the situation was, but he still needed to be gone. Something else would be more important before the week was over. He went back to his desk, grabbed his phone with one hand to call a friend to do some surveillance. The other hand grabbed Ron's paper threw it in the trash.

CHAPTER 14

27 OCTOBER
2E800
PENTAGON
WASHINGTON DC

Tony Parkman had been silent all day. By 5:00, Ron couldn't take it anymore. "Tony, I guess I owe you an apology. I'm sorry."

"No problem."

Ron knew better than that. "It is a problem. Every day I come here, you say something about how screwed up this place is. I said something to someone that just might be able to change that. And I'm getting attitude from you? I don't get it."

"Attitude? Attitude?" Tony was visibly pissed. "All it is, is talk! I'm just talking, Ron! It doesn't mean anything. I know my place here."

"Your place here?"

"Yeah, my place! I'm just another Major that does what he's told. I serve my sentence of three years, two if I'm lucky. Then I move on. By coming here I know I've done my time and the only way I'll ever have to step in this building again is if I'm one of the chosen ones. One of the one percent that sticks it out because they can compartmentalize all the bullshit. The ones that know there is a war on, but are here in this puzzle palace fighting our own war." He was shaking his head. "I hate this job and I hate what it's made me become."

Ron knew exactly how Tony felt. "So you hate me too, now?"

Tony closed his eyes and put his head down. "No." He turned and looked at Ron. "I'm just mad at you because you have balls and I don't."

Ron tried to smile, but couldn't. "You said something to Kusman and I basically bailed out when the firing started."

Ron couldn't let Tony get away with feeling sorry for himself. "You're forgetting something, Major Parkman. You're going to be Lieutenant Colonel Parkman next year, or the year after. I can say things because I will remain Major Benson." Ron let the comment sink in and watched to make sure Tony understood what he meant.

Ron didn't want to dwell on the meeting with Kusman, quickly moved on and asked, "Did Henderson give you any shit?"

"Oh, hell yeah! He said it was my fault that you ran off at the mouth."

"That's what I'm most sorry about. That Henderson gave you shit for something I did."

"That's the way this works, Ron. You're untouchable. You aren't in the fucking system anymore! No offense, but you ain't getting' promoted, so they can't hold your feet to the fire anymore. You don't count."

Ron grunted. Even though it was true, hearing it was an awakening. "It still hurts, Tony."

"I say again, you're done, Ron. They can't do anything to you! That's why Henderson can still screw me over. You can do damn near anything and they won't give a shit." He turned back to his desk. "That's why I'm pissed. It has nothing to do with you. So if you'll excuse me, I'm gonna sit over here and pound my nuts flat."

There wasn't anything for Ron to say that would make Tony feel better. He grabbed his gym bag and rubbed Tony on the shoulder as he left. Tony Parkman was a good officer. It wasn't his fault he couldn't be honest.

27 OCTOBER
PENTAGON GYMNASIUM
WASHINGTON DC

As Ron worked out, he was angry as he replayed the exchange with Tony. He wondered who's fault was it that he was "done" and Tony Parkman got shit on for Ron's transgression? Was it Henderson? No. Was it McCoy?

Even though he was a General, he was only a cog in the machine. Majors were the track that Colonels' ran over to get to the high ground. Generals were the high ground. It was amazing that the high ground was never high enough. He increased the tempo as he jumped rope. The anger was a tremendous stimulant.

An hour later, the fatigue had set in and Ron's mind had mellowed allowing clarity. It wasn't any persons fault. It was just the machine. The Department of Defense machine. Where humans were the fuel that fed the machine and kept it running. Money was the lubricant. One thing Ron knew about machines. You can have all the fuel you want, but it's the lubrication that keeps them working.

CHAPTER 15

3 NOVEMBER
"FINNEGAN'S PUB"
FALLS CHURCH, VIRGINIA

Clyde had given Ron his list of appointments on Monday night. He only had one new client and he was scheduled to meet her Wednesday night, 9 o'clock at "Finnegan's Pub". Finnegan's was more of an old-fashioned 'Speak-easy' than bar. Lots of couples and smaller groups of friends met there to have a drink or two.

Ron made sure he was fifteen minutes early for new customers. He liked to scope out the bar or restaurant and try to figure out which woman was the 'newby' that was in search of an escort. He sat in a booth in the back of the bar in a good enough position so he could watch the door to see his next client, "Grace", before she saw him.

At about 8:55, he nearly spit his wine out when he saw Dorothy Henderson walk through the door. He noticed that she was extremely well dressed. What the hell was his boss's wife doing at Finnegan's? Like a light switch illuminating a bulb, he answered his own question. He quickly slid further into his booth and got the waiters attention. "Could you ask the lady that just came in if she's looking for Steve? If she is, tell him he is going to be late and to just take a seat at the bar, okay?" The waiter grabbed the ten-dollar bill with a smile and nearly sprinted to Dorothy Henderson.

Ron watched as she listened to the question, nodded, and headed to the bar. She took off her coat and revealed a low cut red satin dress that fit her like a glove. She wasn't one of the typical trophy wives a Pentagon warrior would put on parade, but at one time, she turned an awful lot of heads. Dorothy Henderson was trying to re-capture those glances.

Ron had only met her once before. He wasn't even sure if she knew his name. He took a big drink of wine and left another ten on the table. It was time for 'Steve' to go meet 'Grace'.

Ron walked over to the bar and stood right next to Dorothy and ordered a drink. "Could I get a Cab Sav, preferably from Australia?" The bartender headed off.

He turned and pretended to be surprised. "Excuse me! Are you Dorothy Henderson?"

She flushed red and looked around to see if anyone was watching them. Seeing the coast was clear, "Yes. I am."

Ron said, "You don't remember me, do you?"

She looked at him carefully. His hair was longer and he had lost about ten pounds, but she recognized him. It was his name she was struggling with. Colonel's wives sometimes don't remember Majors' names. It would only take her a moment to figure out who he was.

"I won't test you, Ma'am. I'm Major Benson! I work for your husband."

Again, her face flushed red. "Oh yes! I remember you, Major Benson. It's nice to see you again." She looked around the bar quickly and then back at Ron.

He sat down in the chair next to her and smiled broadly. He could tell by her facial expression she didn't want to be seen by anyone she knew and definitely didn't want him to sit down. "What are you drinking?"

"Um, I'll have a Cosmopolitan, please."

Ron got the bartenders attention and ordered her a drink.

"The Colonel is in Kansas this week isn't he?"

"Yes, I believe that's where he is." She searched the room nervously.

"Are you waiting for someone?"

"Ah." She tried to think of something to say. "Not really."

Ron sat there quietly and took a small drink of his wine. He looked at her reflection in the mirror behind the bar. The sleeveless red dress, her hair made up and just enough make up to take off ten years of military wife life.

For a moment, he thought about walking out. Just pay the tab and walk away. She would be none the wiser. He looked at his watch. Ten minutes after nine. It would have been too easy to leave. Yet his conscience would not let him forget his boss's wife was looking for a lover and some semblance of integrity wouldn't permit him to sleep with her. Ron was damned if he left and damned if he stayed. He decided to take the valorous route. He wouldn't leave her sitting alone.

He turned to face her and waited until she looked at him. Underneath her nervous exterior was a very pretty woman. He smiled at her and said, "Steve's not going to be here tonight."

She flushed bright red once again and her mouth opened slightly. She was clearly embarrassed. "You know Steve?"

Ron nodded and looked at her. "I am Steve, Dorothy."

"Oh, my God!" She quickly turned away and faced the bar. Ron could see her looking at him in disbelief in the mirror. "You're . . . Steve?"

Ron nodded and toasted her reflection in the mirror. "Yes, Ma'am." He took a big drink.

"Well." She grabbed her own drink and toasted him back. "Isn't this . . . awkward."

"I was thinking of a different term, but awkward works." He exhaled heavily and looked at her. "I must tell you, you certainly did a terrific job of getting ready to meet Steve." He could see she was embarrassed. "And as much as I would like to be Steve tonight, I don't think that would be in the best interest of either of us."

She lowered her head, ashamed. "No. This would make things much . . ." Her words trailed off. She didn't need to say any more.

Ron was quick to lighten the mood. "Don't get me wrong. I'm serious when I say you look fantastic!" She looked back at him and he could see she was smiling. "But Steve isn't going to be available tonight."

A look of concern came to her face. "You're not going to tell anyone about this, are you?"

Ron smiled broadly. "Who do you think I can tell?" Dorothy Henderson exhaled audibly. Ron continued, "We both have something very secret going on here, Dorothy. You aren't alone in this."

She smiled and turned to Ron as a memory came to her. "You're name isn't Steve?"

Ron took a drink and looked at her in the mirror behind the bar. "Not at work." She nodded her understanding. "It's Major Ronald Benson." He turned to her, pulled his hand up and gave her a mock salute. "Tonight you can just call me Ron."

"At work," her eyes narrowed. "You're the one that's giving my husband an ulcer. He's been complaining about someone named Major Benson when he comes home."

"That would be me." Ron swirled his wine and looked into the glass. "If Colonel Henderson has an ulcer, that would be his fault. I have nothing to do with his health."

"He's headed to an ulcer over you."

"He's headed for an ulcer because he's too tied up to being a Colonel in the Pentagon. I didn't put him there."

"But you're not making it easy for him."

Ron started to get angry. "How did this get about me? Why are you here? Your husband is out of town and you're here. But you're giving me the third degree because he's pissed at me at work? What's wrong with this picture?"

She frowned and looked down at her drink. "I'm sorry. I wasn't being fair."

He exhaled loudly and put his head down. It was him not being fair now. "Apology accepted." Ron relaxed a bit. He was getting too worked up about something that he was not in the middle of.

She was quiet for a moment and stared straight ahead. She took a sip of her drink and said, "We haven't been sleeping together for over a year now." Ron put his drink down and looked at her. "I don't think he's seeing another woman, but I know he's not been with me." She wiped away a tear. "I've never done this before. Honestly, this is the first time!"

"I believe you. I can tell."

She smiled. "I've been thinking about it for a couple months and finally got the courage to find someone, like Steve." She took a long drink and ordered another.

"Why are you looking for Steve?" At that point Dorothy Henderson broke down a personal wall. She let go with years of pent up frustration. Not just sexual, but mental and emotional. The pressure of being a military wife was putting untold strain on her. She resented it. That strain had manifested itself in the Henderson bedroom. She had slowed down so much of her sexual life that her husband didn't even try anymore.

She was confused, she was depressed and she was alone. She had no "girlfriends" since they moved to DC. Some acquaintances, but not real friends. Ron saw she was quick to wipe away another tear. He handed her a napkin and ordered a Diet Coke.

Ron explained to her what her husband did. How being in the military was about being a 'warrior'. The job they currently filled was nothing like being a warrior. He explained that they were probably the equivalent of military eunuchs. That made her laugh and she smiled again.

He told her how he had hurt his wife and that caused them to get the divorce. That led to another ten-minute discussion on marriage and the military. By the time they finished on that topic, Dorothy Henderson was getting a buzz.

Ron said, "Maybe meeting me here is just a sign that this wasn't supposed to happen for you."

She turned toward Ron and leaned against the bar. The drinks were definitely taking effect. "That's too bad. I kind of wish it would happen." She was staring at Ron's face.

"That's probably just liquid courage now. You're on number three, so you're a bit more relaxed."

"I'm more than relaxed. I'm horny." She covered her mouth as she laughed.

Ron rubbed her back. "Some other place, some other time, we could make this happen. I don't think, under the circumstances, this would help you." Ron added, "In the long run, I'm sure it wouldn't help me!"

She laughed again. "How about a short run then?" Ron smiled and shook his head no. She patted his leg and said, "You're right. I know that."

"I'm gonna go."

She exhaled loudly and smiled at him. "Okay."

Ron stood up and leaned close to her. "Talk to him the way you talked to me. You need to try and see if you can work it out. At one time, you two were in love with each other. Try to find that." He kissed her on the cheek. "If you can't," he gave her a devilish smile, "then you call Steve again, okay? Cause you are a damn fine looking woman!"

She smiled a broad happy smile at him. "Do you need a ride, Mrs. Henderson?"

"I'm going to be all right, Major Benson. Thank you."

He stuck out his hand. "Until we meet again, Ma'am." She placed her hand in his and he kissed it gently.

As Ron left the bar, he didn't notice the man sitting near the exit with the military haircut, drinking a coke. There was no way he would have seen his camera.

3 NOVEMBER
313 WASHINGTON AVE
APARTMENT 24
FAIRFAX, VIRGINIA

Alone in his apartment that night, Ron Benson lay awake contemplating the nights event. What could have been a truly disastrous situation, turned out better than he had expected. Dorothy Henderson was a nice woman who was hurting inside. He hoped that by talking about her problems, maybe he had helped her. It surely was better than if he had slept with her. Not only would it have been bad for her, it would have only compounded his already precarious situation at work.

He turned out the light and fell quickly asleep. Everything had turned out for the best. So he thought.

CHAPTER 16

22 NOVEMBER
C RING - 117
PENTAGON

Lieutenant Colonel Janet Stockman was smiling at Ron as she walked by. She was an Air Force Officer that had been around the Pentagon, but was new to the Public Affairs office. Single, attractive and available, under different circumstances, Ron would have asked her out. As it was, he had no time available for his new co-worker.

"You're looking awful happy? Are you losing weight?" If Ron didn't know better, he could have sworn Janet Stockman was trying to hit on him.

"Maybe just a little." He smiled back at her.

"Whatever you're doing, you need to keep it up." She turned and continued down the hallway.

Tony Parkman turned and looked at him. "What the fuck was that?"

"What was what?" smiled Ron.

"You got something going on I don't know about?"

"Absolutely nothing, Tony."

Tony raised an eyebrow. He looked around to make sure some imaginary person was not listening. "I've heard there's a male prostitute in the building."

Ron nearly shit his pants. He recovered by looking around in mock surprise. Ron looked directly into Tony's eyes and pulled him close. He eyed his friend cautiously before saying, "Tony, I'm a male whore and I'm having so much sex that my pheromones are exploding. Women can't help but find me attractive."

Tony stared hard at Ron. For a second, Ron thought Tony really believed the truth. Suddenly Tony burst out laughing. "Holy shit, man!

You had me going! Male whore! You?" He burst out laughing again. "Pheromones!" He laughed again and smacked Ron on the back. "You're killing me!"

For that short fraction of a second, Ron had thought the jig was up. Truth be told, it was an outrageous thought that any officer, especially an officer in the Pentagon was a gigolo. Tony Parkman was absolutely positive his friend was telling him a joke. Ron tucked away the fact that there was a 'rumor' running around the Pentagon. The building functioned on rumors. He filed away the "RUMINT" category. RUMINT was "rumor intelligence" that was so prevalent in the hallways of the Pentagon. Dozens of little fictitious factoids floated around the building that were started over coffee or overheard partial conversations. Ron wanted to change the subject before Tony asked any more questions.

Tony was only too quick to help him out. "Maybe you should ask her to the Ball?"

"I'm going alone. I don't have a date. I'm not looking for a date. And frankly, if Henderson weren't demanding we be there, I wouldn't go."

"You can dance with Cathy then." Cathy was Tony's wife. "I hate getting dressed up and I dance like I'm from a long line of white men."

"Tony, you are from a long line of white men." His friend smiled at the crack. "I'm just going to make a token appearance and leave early. Speaking of which, I'm leaving now to go to the gym. I want to be ready to get out of here tomorrow with the Thanksgiving holiday coming up." Ron smacked Tony on the arm with his paperwork and headed back to his cubicle to get his gear. Ron had a huge smile on his face as he yelled, "Later, Power Point Pecker!"

Ron knew all too well about the Microsoft 'gods'. Everyday as he flashed by cubicles, he saw the 'serfs', planting their own virtual crops and sending electronic prayers. Not with hoes or seeds, but with tools of the demon Gates. Everywhere he looked, his fellow staff officers were on computers planting seeds and sending prayers to their blasphemous gods. Word; the god of administration, Powerpoint; the god of operations, and Excel; the god of logistics. Ron knew the god of information was constantly in need of sacrifice. He had his own special alter in the Five-sided wind tunnel.

Tony yelled back, "Hey! Let's not let anybody know I even have a clue what Power Point is! Bill Gates is a traitor!"

Tony watched his friend as he walked away. He had noticed Ron had lost weight. He looked at his own stomach and pinched it. He yelled after Ron, "Can I come with you? I've got staff officer disease starting. I have strived to achieve the minimum standard for physical appearance and have attained it!"

Ron just waved the papers over his shoulder and kept walking. The mid-career officer's gut would have to be Tony's prize. Major Benson was becoming addicted to working out again.

On his way to the gym he just happened to pass by a cubicle with two occupants raising their voices. Both officers were lieutenant colonel's, one was Air Force and the other in the Army. They were in a heated debate about a presentation they were to give.

Ron stopped for a moment and overheard, "This font is all fucked up! The standard is Arial 12. I'm not sending it up like that!"

"There isn't a damn thing wrong with that! New Times Roman is more appealing on the eye than fucking Arial! And besides that, he's an old man and he can't even read 12 font! You gotta make it bigger!"

"I'm not touchin' it! You fuckin' do it!"

Ron shook his head and looked at his watch. *The God's were at work again.* It was after 5:30 and the pressure meter was sounding off on any and all officers that had a deadline to push paper to their boss. *Just one more tiny sacrifice to the insatiable God's and their information war.*

As he stepped out side he felt the chill in the air. He looked over his shoulder at the building behind him. It was the building that changed warriors from men fighting for their country against all enemies, to men fighting each other over the font size on a piece of paper.

Ron picked up his pace as he headed to the gym. He no longer wanted to work out. The fact that Tony had even heard a rumor about a gigolo in the Pentagon was disturbing. It gnawed away at Ron, like a crappy song he heard that he couldn't stop humming. The beat was all right, but the lyric, 'gigolo' kept repeating itself. He needed to work out to clear the Pentagon from his mind.

CHAPTER 17

23 NOVEMBER
2B800
PENTAGON

Ron stayed late at the PA office to try to finish another presentation. As he sat there alone in the office, he couldn't focus on the task. The slides were not his work. They were Henderson's and they reflected what his boss felt. Unable to convince himself to share his boss's perspective, he did the minimum. Ron made the fonts all the same; less one of the argumentative lieutenant colonel's miraculously appear to start bitching at him about Pentagon standards. He didn't touch any content, made no deletions or additions to a single slide. His heart wasn't in the work.

His head was drifting off thinking about what adventure was waiting for him on Clyde's list. He couldn't care less about Henderson's slides. They were distracting him from where his heart really was. Ron shook his head. He was having problems doing his own work, much less his bosses. He solemnly turned off the computer, turned out the lights and found himself nearly sprinting out the door.

It was the Tuesday night before the Thanksgiving holiday and Ron had made no plans. He hoped that Clyde's list would give him something to do over the long weekend. The Pentagon only made him depressed. He wasn't a warrior anymore. Hell he wasn't even a good staff officer. Yet he was still finding reasons to smile. His new pastime, or paying hobby as he considered it, was managing to take away his fatigue. He was looking forward to working on the hobby over the holiday.

As he left the five-sided big top, Ron Benson pulled his winter coat tighter around his neck. The cold air made his breath appear as he walked.

The sky was dark and the sun was already below the horizon. The gray day was turning into a cold, gray night. The same way his life was turning from gray into black. Ron shook his head and tried to smile. He was no longer worried about what he did at the Pentagon. It was no longer a duty. His Pentagon life was just a job.

It was time to meet Clyde. He'd hoped Clyde would cheer him up. Clyde was always good for that.

23 NOVEMBER
MCKINSTRY'S STEAKHOUSE
ALEXANDRIA, VIRGINIA

"I guess the worse thing is that I pissed off the only real friend I had. I think we're okay now. But the whole thing hurt him. I suppose I lost his trust."

Clyde was sympathetic to Ron's plight. "I don't know what to tell you brother." He put a big piece of the New York strip in his mouth. That didn't keep him from speaking. "You should just pack it up. Say the hell with it. You can come work for me full time." Finally, Clyde found his napkin and wiped his chin.

"It's just not that simple." Ron looked at the surrounding crowd. There were no military folks at the restaurant. He exhaled loudly. "There is a piece of me. A really large piece that loves being an officer." He searched for the right words. "I want to be a warrior again."

Clyde nodded. "You want to be in Iraq. You want to be a leader of men!" Clyde grabbed his glass of wine and made a gesture of toast. "Better you than me.

"I'm sure some psychologist has a syndrome or is getting paid for a study that explains why old farts like you want to go to war. You need to look in the mirror, my friend. War is a young mans game."

Ron shook his head disagreeing with his new mentor. "Remember that old saying, 'Old enough to fuck, old enough to fight'?" Clyde nodded and showed his wide grin. "That still applies to me. I'm still young enough to," he smiled as he tried not to be crass, "take care of business, both in a bed and in battle. But I gotta wonder, Clyde, am I doing this as an outlet because I can't go to war?"

Clyde pondered the question. "I can't say for sure Ron. I highly recommend that you stop getting so deep in your personal psychosis, and start enjoying your part time employment. You don't know how long you can keep doing this, but you know you're your chances to lead soldiers gets more remote every day. You need to focus on the now." Clyde pulled out a piece of paper with a list of contacts for Ron.

Ron grabbed the sheet and tucked it in his pocket. "You're right. As usual." Ron grabbed his wine glass and took a small drink. "I need to stop thinking so much and enjoy this." He raised his glass to Clyde. "At least until the long tour to Leavenworth takes it all away."

They talked for another 30 minutes, but Ron found himself drifting between Clyde's carefree attitude and the thought that he was a criminal. He knew what was bothering him. He missed leading soldiers. Whether he was sent to the Federal Penitentiary in Kansas or suffered as a staff officer for the rest of his career, Major Ron Benson would never lead troops again.

CHAPTER 18

28 NOVEMBER

CHESAPEAKE MARINA AND YACHT CLUB

CARLISLE ON CHESAPEAKE, MARYLAND

There was one name on the list that Ron was glad to see again. What he was confused about was where he was supposed to meet her. The agents on the dock were almost glad to see 'Steve'. At least one was happy to see him.

"Hello, gentlemen."

"Good evening, Mr. Jones," said Rollie Gault.

"It's good to see you guys again."

"Can't say the same for me, Mr. Jones," said agent Klein.

Ron just smiled. "Good to see you, Agent Klein."

It was time to get down to business. Klein approached Ron and motioned for him to put his hands up. "Excuse me, sir, but I need to check and make sure you are unarmed."

Ron did as he was told and looked over Gault's shoulder at the yacht. "Nice freakin' boat!" Gault lifted up Ron's jacket from behind and patted him down.

Klein said flatly. "It's her husbands', Mr. Jones."

Gault cleared Ron and got close to his ear. "Don't pay him any mind. He doesn't much care for what you do or this job, but I kind of enjoy you coming around."

"Why is that?"

"Usually after she sees you," he looked around to make sure no one else was listening, "she's pretty nice for three or four days." Ron smiled broadly. "After she see's you, she isn't nearly as bitchy."

"I'll see if I can still do that again tonight, Agent Gault."

"Please do, sir. You're clear to enter the boat."

Ron nodded and climbed on the yacht. He was glad the bay was calm. The yacht was big enough to require a gangplank and was moored in three spots. He turned quickly to ask Gault a question. The agent read his mind.

"There is no one else on board, sir."

Ron turned and looked at the boat. It was at least 70 feet from bow to stern. The most impressive thing Ron noticed was how clean the boat was. For an instant, he thought about taking off his shoes, but the 45 degree temperature was continuing to drop. The shoes would have to stay on. At least for a few more minutes.

Ron entered the living quarters of the yacht. It was like a museum. Fine crystal vases and silver decorations were everywhere in the room. There was a painting of a woman behind the bar. As Ron got closer he could see it was Denise. It was probably painted ten years earlier, but Ron could tell it was her. She was beautiful.

On the table in the corner was a bottle of something setting on ice. There were two crystal glasses sitting next to the silver canter. There was a huge vase full of flowers on a coffee table across from what was the largest couch Ron had ever seen. The flowers were fresh and filled the room the smell of springtime. He knelt down to smell them and for a moment, forgot why he was there.

"Good evening, Steve." Ron turned quickly as she had surprised him. She was dressed in a long black, sleeveless evening dress with a plunging neckline that went down to her belly button. Ron noticed it fit her just a little too tightly, but that was congressional stress showing. Five years ago, the Senator probably filled the dress perfectly. She had made an all out effort to look her best and it showed.

"After the pajamas, I thought I needed to try something a little different this time." Ron could only smile. "Did I do all right?"

Ron stepped close to her, smiling broadly, he took her hand softly and pulled it to his lips. "You look beautiful." He softly kissed her hand and looked into her eyes. She blushed and smiled. She was instantly ten years younger. Ron was only wearing a sports jacket, over a turtleneck and a pair of Dockers slacks. "It's my turn to be under-dressed."

She laughed. "I guess you can say I wanted to make sure you know I don't always run around in pajamas."

"I didn't think you did. I mean, I've seen you on C-SPAN and granted, you dress conservatively for that. But tonight," he shook his head, "I guess the word 'stunning' comes to mind."

The Senator smiled a huge smile. Then her expression changed as something seemed to dawn on her. "All the women you make love to? How am I supposed to know if you're just bullshitting me?"

Ron understood her perspective, but didn't want to tell her that there really weren't 'all' those women. He went to the canter and filled two glasses, handing her the first one. "You look beautiful. If this yacht were filled with a hundred women, none would compare."

Again she blushed. Ron touched his glass to hers and took a drink. He decided to take it slow. He would be there through the night so he was in no hurry. He would let her desires set the pace. As she moved to the couch, he did notice she was not wearing any shoes. She lay back against the pillows and stretched her feet on the couch.

Ron sat next to her outstretched feet. He put his drink down and grabbed a foot. He slowly started to massage her foot. The Senator closed her eyes and laid her head back against a pillow. "You can stay forever if you're going to do that."

Ron smiled, but changed directions on her. "Where'd you grow up?"

Her head snapped upright, "What do you care?"

"Just making conversation. I doubt you want to take about the Armed Service Committee meeting." He pushed his thumb slowly into her heel.

She inhaled sharply with pleasure. "You're right." Her head popped back against the pillow and she began talking about her child hood. She had grown up in Indiana. Daughter of two schoolteachers. She never participated in sports because she thought she was uncoordinated. Her parents scraped every dime they had together to put her and her brother through college. It wasn't until her attendance at the University of Indiana that she started to care about politics. She moved to Maryland and worked some mediocre assistant jobs in the state legislature and on staff of a congressman from Baltimore. The sense of power got to her so she decided to try running herself.

To Ron's surprise, she originally didn't want to run as a Democrat, but couldn't count on getting funding from Republicans because she was a woman. That made her choice easier. It was during her campaigning that she met her husband at a fund-raiser. Ron switched to the other foot.

She titled her head to look at him and said, "I know what you're thinking."

"I wasn't thinking anything." He pushed on the balls of her foot and she grabbed the back of the couch and groaned with pleasure.

She exhaled loudly. "You were thinking 'that's where she met the rich old man that she took to the cleaners', weren't you?"

He smiled and moved his hands toward her toes. "Nope."

Whether it was the drink or the massage, she did something she would never usually do. She dropped her reserved side and confided in him. "I was in love with David Samuelson." She took a long drink and set her glass down. "Initially, I believe he was totally in love with me. Eventually, I think it was the power that I was gaining."

"And you with him? I mean, initially?"

"Of course. I didn't have time for anyone else so he had no competition." She looked up at Ron. "It wasn't until years later that I found out I was merely a trophy for him to display." She became thoughtful. "I suppose you could just say we have drifted apart." She lay back again. "Now we rarely talk."

Ron put her feet back towards the couch and slid around towards her. He lay next to her and softly touched her face. He slowly moved his lips to hers and kissed them, ever so lightly. He pulled back and looked into her eyes, "If you want to talk more, I'll listen."

She reached up and held his face. "I think I'm done talking for the night." She kissed him, softly at first. Eventually the passion rose in her and she was taken away.

Ron thought about carrying her to the bed, but between the two of them, the lust couldn't wait. He was well aware that he wasn't making love to her as much as he was being seduced. He didn't care for her politics, but he was taken away by her love-making. She was passionate and lustful, tender yet aggressive.

Perhaps what was the most remarkable was how she was able to focus, just on the two of them. He was in total awe of how she could compartmentalize what she was, one of the most powerful women in the world, and separate that from the woman she wanted to be. If was difficult for Ron to fathom that the woman he watched on television was the same lovely and driven lover that he had come to know.

The bay was extremely calm that night, but the two security agents couldn't help but hear the noises coming from inside the boat. The occasional groans and moans broke the silence of the night breaking the monotony of what could have been a boring night.

Gault looked over and said, "I gotta take a vacation and get my wife on the love boat."

Klein was shaking his head. He was listening to the loud, rhythmic knocking coming from the cabin below, accompanied by more moans. "You're sure she's all right? Should we go check on them?"

Gault took a big drag on his cigarette. "Naw. I'd say she's more than all right." He looked down at his watch. "I just hope they go to sleep soon. It's getting chilly out here." Klein turned and said, "It's goin on midnight. They should be done by now!" Then something else dawned n him. He shook his head. "It's kind of making me horny!"

"Hey, the longer he goes, the better off we are! If this helps her relax, that's fine with me." He looked out across the bay. "Last time she was with this guy, she was a queen for about three days. If he can make that woman come back and," he lowered his voice, "get rid of the bitch we normally deal with, I hope he goes all night."

"Oh, yeah!" Klein agreed with the comment. The noises continued to come from the cabin below.

After about five minutes, Klein asked, "How much do those cruises cost?"

CHAPTER 19

28 NOVEMBER
CHESAPEAKE MARINA AND YACHT CLUB
CARLISLE ON CHESAPEAKE, MARYLAND

The digital clock read six forty-five when Ron rolled over. He had been sleeping soundly and it took him a moment to figure out why the bed seemed like it was floating. For an instant, he became nervous. Then he remembered where he was and the fact that the whole room was floating and he laid back on the pillow.

Denise was already in the tiny galley making some scrambled eggs and toast. A small TV was playing CNN news on the counter as she moved around the kitchen. Ron wasn't certain if she was a morning person or not, so he announced himself then sat in a chair to give her some space. He need not have worried.

She was a bundle of energy as she cooked. "I've already eaten. I didn't want to wake you up. Is scrambled, OK?"

"That's terrific."

"I have to go shower. I need to get the office for a nine o'clock appointment."

"Not a problem. I have someplace I need to be, too." Ron took a cup of coffee as she handed it to him. He was well aware of her appointment. "You're asking questions to the military Chiefs today, aren't you?"

Denise turned and looked at him with narrowed eyes. "That's an odd question coming from you?"

"You mean your 'prositidude' can't be a concerned citizen?"

Her eyes narrowed further. "Why do you care?"

"My tax dollars pay their salary. You need to make sure they're doing their jobs to defend our country."

She turned to the eggs. "Is there something you need to know from the Joint Chiefs of Staff, Mister Concerned Citizen?"

From the recesses of Ron's slightly awake mind, a light came on. "Why, now that you ask, Senator, do you have a piece of paper?"

She turned and looked at him with a confused glance. "Yes, I do Mr. Jones." She found some note paper and a pen. Ron quickly jotted down three questions. Denise looked at them inquisitively. Ron could see the smile appear on her face.

"These are . . . very interesting Mr. Jones." She looked at him with a huge smile. "I think you are a bit more than just a concerned citizen. These are excellent questions that I would like to know the answers to myself."

"I don't spend all day lying around in bed with women, Senator."

She looked at the questions again. "Strategic Communications, huh? Maddrasah's? What do you know about Madrassahs?"

"I'm a news junkie. I heard there are bad things going on at those so called, schools. Just wondering what the policy was? I mean, if I was a soldier, wouldn't those questions be things I would want to know?"

She nodded. "Yes." She looked at her watch and frowned. "Damn, I gotta go." She bent over and kissed him. "Thanks for these." She started to leave, but came back and kissed him again, a devilish look in her eyes. "And for last night! See you soon, I hope."

He lightly touched her arm. "Me, too! One other thing." She stopped in place and looked at him. "Be nice to your security guards."

He could see it took a second, but she understood the request. She nodded, "OK. I can do that."

It was eight o'clock when he finally surfaced from the cabin. It was a chilly, yet bright day on the Chesapeake Bay. Gault was still there. "Good morning, Agent Gault. How are you today?"

He was smiling. "I'm good Mr. Jones. A little bit tired, a little cold, but good."

Ron jumped onto the dock and stuck his hand out to the agent. Gault thought that an odd action from the man that had just committed adultery with his Senator. But he took the hand and smiled. "Sorry you had to stay out here in the cold."

"That's all right, Mr. Jones. It appears the Senator is in a very good mood this morning."

Ron smiled and said, "Whatever I can do to serve the country, Agent Gault."

"Tough duty, Mr. Jones."

"It ain't Iraq." Gault nodded. "Just call me if she gets too grumpy again. Take care."

"You take care, too!" Gault was amused. Jones, or whatever his name was, seemed like a nice enough guy. He actually reminded him of one of the officers' he had met up on the hill. Out of habit, he followed the man as he walked to the parking lot. He couldn't make out the license plate number, but he did notice the blue military sticker on the windshield. Whether it was habit or curiosity, he wrote down the make and model of the vehicle. He had a broad grin as he was doing it. "It can't be," he said to no one. "I'll be damned."

CHAPTER 20

30 NOVEMBER

2B800

PUBLIC AFFAIRS OFFICE

PENTAGON

Ron hated being called in to H2No's office. The feeling was exacerbated by the fact that it was 15:45 and most of the office had left. As he waited to enter, he listened to the commentary from some reporters on CNN discussing how the Senate Armed Services Committee had put the Joint Chiefs on the spot earlier in the day. Apparently, the General's didn't give adequate answers to some of the senator's questions. Ron was smiling when he was called into Henderson's office.

Henderson was the exact opposite. He was as pissed off as Ron had ever seen him. He directed Ron to have a seat. This didn't sit well with Ron because he had no desire to be there long enough to need a seat. It didn't matter because Henderson had no intention of keeping him there long at all. Just long enough to make sure the shit had rolled down hill far enough to get off his desk and into Ron's inbox.

"I've been getting crapped on all afternoon from Generals around the building with questions trying to respond to the damn Senate Armed Services Committee questions." He turned his computer screen so Ron could look at it. As if on queue, another email popped into his mail account.

"Look at this! Look at this! I've got," he squinted as he looked into the screen, "212 emails! What a crock of shit!" He pulled the screen back so he could see it clearly and got to his point. "I've sent you this email." He handed Ron a piece of paper that had a half a dozen questions and a summary paragraph on it. "I need you to prepare me point papers or white papers on those four boldly lettered items."

Ron looked at the email. He kept his smile to himself. "When do you need them, sir?"

Henderson looked confused. He had expected some form of bitch from his subordinate. When none came, he was actually pleased. "Well if possible, tomorrow." Again he waited for some form of protest.

Ron read over the questions quickly. "I don't really have any plans for this evening." He lied. "I think I can get at least three done and I'll work on that last one tomorrow morning, if that's all right, sir?"

Henderson's attitude changed immediately. "That would be fantastic." The thought that Henderson may have actually worked on one himself never crossed the Colonel's mind. He was actually smiling when he dismissed Ron. Ron stood up, saluted and left without any further guidance.

He still had the email in his hand when he sat down at his desk. Tony Parkman had already left. It was after four o'clock and there were only a couple people left working. Ron put his feet up on his desk and re-read the email. He already had three of the four questions answered. They weren't in the format that Henderson wanted, but that was easily fixed. The fourth question; "what advantages are there to active Public affairs vice passive", was one he had not suggested to Senator Mitchell-Samuelson. But it was a great question and would take a few minutes to answer properly.

Ron threw the paper on his desk and placed his hands on his head. He would have plenty of time to get ready for his nine o'clock appointment.

5 DECEMBER
JEFFERSON MONUMENT
WASHINGTON, DC

"This is an absolutely beautiful piece of architecture, don't you think?"

Ron looked up at the building and blew into his hands to try to get them warm. "I'll agree with that." He looked around. "I've never really gotten this close to it."

Clyde walked down the steps, faced Ron and handed him a list with eight clients names and phone numbers. Clyde said, "I don't believe the founding fathers ever had the vision of modern day Washington in mind when they started this country."

"I imagine they would be rolling over in disgust." Ron was unable to contain his feelings.

"I don't think they ever believed that money would be the main influencer in politics. I imagine their concept of power was more geographic, than fiscal. They started us off as a Republic, ya know? They wanted the power at the state level, not the federal level. The federal government was only supposed to take care of mutual defense and allow states to have the free trade so everyone prospered. Now we have an unimaginable amount of money corrupting a completely fantastic concept like our democratically elected government."

Clyde was in a professorial mood. "You know when the Iraqi's voted, they estimate that over 90 percent of the eligible population voted. That's with the threat of suicide bombers at the polling places, too!" He looked at the monument. "We barely get 40 percent turnout."

Ron chuckled. "What are you? Nostalgic for the 18th Century professor?"

"Who me?" Clyde's smile beamed. "Hell no! I'm just a taxpayer!"

Ron laughed out loud. "I know you're not paying taxes on all your income! Hell you can't even tell Uncle Sam where your income comes from!"

Clyde smiled. "As far as they know, I'm a successful gambler."

Ron shook his head and smiled. "Yeah, you're a gambler all right." They started walking back toward their cars. The night air was turning cold. The flurries would start any minute.

Ron began thinking of his own financial situation. "I sent my daughter twelve thousand dollars."

Clyde was impressed. "Damn! How did the Ex feel about that?"

"What do you mean?"

"I mean, you just don't send a check to your daughter, and not send any money to your ex-wife."

Ron thought for a moment. "I don't think she will have any problem with it." He started to ask another question, but put the thought away.

"All that extra money you're making, you need to start planning on what to do with it."

"Should I be worried that I have that kind of extra money now?"

"Hell, no! You should be thankful."

"Oh, I am. I just wonder what the IRS will think?"

"I don't think you should tell them. You claim your Army pay and keep the rest in banks outside the US. If you are starting to make that kind of money, you should start diversifying. I'll send you an email with some options."

They walked a little further and the snowflakes appeared. "I have that stupid Presidents Ball this Saturday."

"The President's Ball? Did you get invited by a certain client?"

"Oh, no! As a Public Affairs officer, I'm required to attend. Not just Henderson, but the General and the Assistant Secretary want the whole team there. That being said, we are all going to attend. So I am unavailable for Saturday."

"I'll clear your schedule. Now that I'm your secretary." Ron laughed. "That should be a very entertaining night."

Ron looked up at the snow and squinted. "I don't know how entertaining it will be. I intend to make my mandatory attendance known and disappear before anyone misses me."

It sounded so simple when he said it and in theory, it was a great idea. But just like a war plan, it never survives first contact.

CHAPTER 21

12 DECEMBER
THE WHITE HOUSE
WASHINGTON, DC

Major Ron Benson was doing his best to be a wallflower. He was wearing his Army dress mess uniform complete with miniature medals. It must have been all the recent trips to the gym, because the uniform fit him like a glove. He was looking like a poster child for Army recruiting.

Ron took a glass of white wine from a waiter as he walked by and slowly strolled around the room. The White House was a gorgeous place for a ball. The President's staff held nothing back and presented the "People's House" as a beautiful tribute to the season of Christmas. The beauty was equal to any Royal Palace around the globe.

A few of his co-workers and their spouses stopped by to exchange holiday greetings but quickly were on their way to visit with others. Ron did his best to stay away from celebrity visitors, members of congress and General Officers. Given the opportunity to choose, any waiter with alcohol was a better option than conversation with one of the 'chosen ones'.

Of course his attempt at solitude couldn't last forever. Lt. Col. Jan Stockman came over to him with a broad smile. She was looking positively beautiful in her Air Force dress blues. Ron performed the customary glance of medals that is an unwritten law for officers. He subconsciously noticed the metals she was wearing and weighed them against the medals she wasn't wearing. Her Joint Meritorious Service Medal was appropriate, but Ron was curious about her Legion of Merit. The 'LOM' was generally reserved for retiring officers or those headed to General rank. The display indicated that she was in the running for a General's star. Although extremely attractive, single and an obvious player in military circles, Ron considered her a friend and nothing more.

"Major Benson, you seem to be keeping a low profile tonight," exclaimed the best-looking lieutenant colonel in the US Air Force.

"Part of the plan, Ma'am," said Ron with a shy smile.

He noticed she was looking at his medals. She reached over and wiped away an imaginary piece of lint from his chest. "Bronze Star, Major Benson?"

"Company Commander in Somalia, Ma'am."

"Ouch!" The response was a double entendre. 'Ouch' because the combat was nasty, lots of soldiers got killed and wounded. And secondarily, it didn't mean anything to the hierarchy at the Pentagon. If you didn't get your awards from Iraq or combat in Afghanistan, your combat awards didn't mean shit. Old fights were passé. It was the current war that meant you were a player. In her eyes, he was yesterday's news. He could've been Brad Pitt with a Navy Cross, but because it was from a battle nearly ten years ago, she wouldn't let him drive her Porsche, much less her.

Before Lt. Col. Stockman could leave, "Hello, Major Benson," came a voice from out of nowhere. It was Dorothy Henderson dressed in a beautiful long green evening dress that hung low on her shoulders. She reached up and gave Ron a kiss on his check. The scene left Colonel Stockman taken aback. She visibly leaned away from the pair is shock at what she was seeing. Ron was very in tune with his surroundings and did not reciprocate. He did gently grab her hand and give it a light squeeze..

Displays of affection are frowned upon in the military. Military men as well as women usually found public displays of affection unwelcome. It came as a bit of a surprise to Ron, but he could see this wasn't the same woman he had met in the bar. Dorothy was genuinely happy.

Lieutenant Colonel Stockman was literally speechless. She didn't think the pair had anything going on in a 'sexual' way, *after all, that's the Colonels' wife*! Yet, they seemed to be more than friends.

Ron stepped back to get a better look at Dorothy Henderson. "How the hell are ya?" Ron was smiling, but the surrounding atmosphere terrified him. He knew deep down he and Dorothy had an unspoken bond, but the military man that he was knew he was way out of bounds by being this familiar to her.

Yet he was unafraid. His 'hobby' had given him confidence. At that particular moment, he made a conscious decision to put the uniform away and be a human. Dorothy was his friend. He was going to treat her like one. Even though the implications lasted well beyond what was left of his career. *I know exactly what I'm doing.* So he thought.

What he didn't know was a man across the room dressed as a waiter was watching his every move. His haircut gave him away as a service member, but he was not there representing the military. His mission was to observe one, Major Ronald Benson. If possible, take pictures of the Major involved in any activity that would indicate nefarious behavior. The man watched Ron and Dorothy. *She was the woman in the bar.* He pulled the miniature camera from his pocket and moved closer. *The General is gonna love this!*

Ron smiled at Dorothy. "I'm guessing you're doing well?"

She nodded with a huge smile. "Oh, yeah. Things are going much better." She pulled Ron close and said, "Everything at home is much better, too." Jan Stockman couldn't quite hear what was said, but understood she wasn't supposed to.

At first, none of them noticed the waiter as walked up holding the tray of drinks. The camera silently snapped away as he approached. As he got within ten feet, he smoothly put the camera away and mustered a smile. "Excuse me, please. Drinks anyone?" The waiter was smiling as he offered champagne.

Ron nodded, took glasses and passed them out. "Thank you." Ron started to turn back to Dorothy, but had a strange feeling about the way the waiter looked at him. He looked too old to be a waiter. Something seemed wrong, but he couldn't figure it out.

Dorothy lightly grabbed his arm and his suspicions went away. He turned to her with a smile and picked up the conversation again.

Dorothy said, "I may have even talked Harold into a vacation this month."

"That's great. I'm glad to hear it." Ron turned to look for the waiter, but he was already gone. Then he noticed something else troubling. Jan Stockman had physically stepped back as if she was no longer associated with the pair. Just when Ron had mentally absorbed what was taking place, he heard another familiar voice behind him. He recognized it, but he wasn't sure he was being addressed. There was a tap on his shoulder. He knew instantly who it was.

"Steve! Steve is that you?" It took Ron a moment to understand what was happening. He turned away from Dorothy Henderson to see Senator Denise Mitchell-Samuelson.

"Why Steve Jones, it has been so long since I've seen you! What are you doing now?" The Senator calmly asserted herself into the threesome. Just out of habit, Denise Mitchell-Samuelson looked down and saw the nametag on the officer's uniform. For an instant, it was her turn to be confused. The nametag read 'Benson'.

She was enough of a politician to know how to recover from her faux pas. She was enough of a woman to know she had been deceived. Her response was immediate. "Oh, my! I'm sorry. I thought you were someone else." She was less than a foot away from Ron.

He knew what he needed to do as well. "Hello, Senator. I think you may have me confused with someone else. My name is Major Ronald Benson."

Her face never even flushed. "Well Hello, Major Ronald Benson. You look an awful lot like someone else I know." Her eyes were piercing into his as she managed to overcome her confusion.

Ron tried to lighten the mood. "A lot of Senators tell me that, Ma'am." He quickly introduced Dorothy and Jan to the Senator. She was polite and political. She also realized she no longer wanted to be anywhere near 'Major Ronald Benson'. Ron looked behind the Senator and noticed her security detail. Gault's expression was stoic. Ron tried to make eye contact, but the agent avoided any glance.

"It truly is a pleasure to meet you all. Thank you for your service to the country." She shook Jan Stockman's hand then looked at Ron. Suddenly it appeared they were the only two people in the group. "Perhaps Major," she was deliberate as she looked down at Ron's nametag, "Benson, you would join me for a dance later?"

Ron felt very small. Although Denise was a client, he liked her. Perhaps too much. This was a woman of class. A woman of stature, elegance and grace. He had fooled her. He had used her as much as she had used him. He tried to salvage any relationship they had. "A moment of your time, Ma'am?" He grabbed her hand and pulled her away from the group. "Excuse me ladies."

A few steps away, he stopped and turned the Senator towards him. He exhaled audibly as he thought about what to say. "You have to understand why I couldn't tell you my real name."

It was obvious she wasn't happy. But she was also a very intelligent lady. "I understand," she paused for a second before adding, "Major Benson."

"I work in the Public Affairs Office at the Pentagon."

"I suppose it's good that I didn't know your real name."

Ron smiled sheepishly. "So much for that." She shook her head and smiled.

"I would still like to take you up on that dance later, if you meant it?"

Senator Denise Mitchell-Samuelson was also in a quandary. She had been fooled. Not just about having slept with a male prostitute, but having one that had an alternative life. A life of prominence. A life of visibility. A life where she would see him again at the White House of all places.

Yet whether he was 'Ron Benson' or 'Steve Jones', he was the only man she had bedded in three years. She had an attachment to him. She looked deeply into his eyes. They were both on shaky ground. "Let's see how the night fairs, Major Benson. If you're still here, and I'm still here, maybe the music will play something we both feel like dancing to."

Ron thought he understood what she said. "I guess I'll check back with you after dinner." He stepped close and grabbed her hand. He pulled it toward his lips and kissed it gently. He looked up at her and said, "I'm sorry, Denise." Then he quickly turned away.

From their vantage point just feet away, Lieutenant Colonel Jan Stockman turned to Dorothy Henderson. "Did you get that? She called him Steve Jones. Those Senators must meet so many military people that they get confused. I understand why she probably can't remember their names. She doesn't even know Ron Benson."

Dorothy Henderson knew differently. "Maybe she knows him better than we do." She turned and nodded at Jan and tried to muster a smile. Then she caught Ron kiss the Senator's hand. *He definitely knows her better than we do.*

Inside she was smiling broadly because she knew and understood more about Senator Denise Mitchell-Samuelson than Jan Stockman would ever know. She knew even though the Senator didn't know Major Ron Benson at all, the Senator knew 'Steve Jones' very well.

After dinner, Ron positioned himself close to agent Gault. He was being diligent in his duty, so Ron was careful as he approached the agent. He cautiously stood beside the agent and said, "How's your night going?"

Gault, always the professional kept his focus on the Senator ten feet in front of him as she talked to a three-star General. "I've had shitty nights before. I don't see this one getting better."

Ron smiled to himself. "Guess I can't get you a beer, eh?"

"Not right now, Major Benson."

Ron paused. "You didn't tell her. I thought you would have to do a background check?"

"Not until the past week, sir." For the first time he turned and looked at Ron. "For something like this, a one time thing is no big deal. After two times, and it became apparent you were becoming a consistent element in her," he searched for "environment. I watched you as you got into your car."

"License plate?"

"That and that big blue sticker in your windshield that shows the world you're an officer were a big giveaway!" Ron shook his head. "That's like a bulletin board."

"But she didn't know my name tonight."

Gault turned back to observe his assignment. "We hadn't anticipated seeing you here. She didn't need to know your real name. It's our job to protect the Senator. We tracked your information, did the background on you and determined you were not a threat." He turned and looked at Ron. "And you make her happy."

Ron nodded and looked at the Senator. She noticed his gaze and stared at him. "Don't know if I'll be able to do that any more, agent Gault."

Gault looked around the crowd searching for terrorists or kidnappers that couldn't get an invitation. "That's a damn shame, sir."

Senator Mitchell-Samuelson had excused herself from the General and walked over to Ron and her protector. Over the speakers, John Lennon was singing about a happy Christmas and that war was over. "Would you like to dance, Major Benson?"

Ron hid his smile. "Sure. I'd love to."

She was quick to grab his hand and start dancing. "I love this song." Ron wasn't sure, but he thought that she was leading as they moved across the dance floor.

"Any song that is about peace and people getting along is a good song," said Ron.

"This coming from a Major in the US Army?"

"Ma'am, nobody digs peace anymore than someone that's fought for it."

"You've fought for peace?"

"Still do it today."

"As a public affairs officer?"

"This war today is as much about public affairs and perception as it is about bombs and bullets." She grunted and moved lightly on her feet.

"They use the internet and television to push bullshit to a receptive audience with immediacy. We write sound bites and allow our own media to provide opposition elitists a platform to voice criticism without reservation. Where are the voices that opposed terrorism and suppression of the innocents?" She turned her eyes to meet his. "Long after we're gone, there are gonna be kids in Iraq that want to be able to choose. Unless America gives them the chance to choose, they will never have the opportunity."

"I understand how you came up with those questions for the Chiefs last week!" Ron tilted his head and winced, "I was hoping you wouldn't remember that."

"Sorry. Having a memory is one of my faults I guess." She decided to ask him a question. "As a Public Affairs Officer, you think what you're doing is helping?"

"It doesn't take bombs and bayonets to win wars. The truth works better in the long run."

She looked into his eyes. "What is the truth Major Benson?"

Ron started to tell here all the things he knew were right about the war. That war was a necessary cleansing. A process of pain that led to healing and eventually the cure. War wasn't the problem. The inability to devote ones self to war was the problem. If the country was collectively opposed to the terrorists in Iraq and Iran and Syrian, the war would be fought as it should be. Fought wholeheartedly, with all freedom loving people supporting the effort. However, Ron was a realist. Real about the war and real about the woman in his arms.

"The truth is," he exhaled slowly, "that nothing matters tonight. It's Christmas. I believe Christ was born to save us. For all our sins. I can't pick up a stone and throw it at anyone that sins." He pulled her closer to him without thinking of the consequences. "I'm sorry I didn't tell you who, or what I am, Denise. It wasn't the truth."

She smiled. "That's all right, Major Ronald Benson. You are not alone in your sin." She tucked her head on his shoulder and danced. They slowly moved to the music. Ron was careful to keep his hands still and visible as he knew people were watching their every move. Across the room, the camera snapped away. The lens captured everything.

As the music ended she reached up and whispered into his ear, "My car will come by outside the gate on Pennsylvania Avenue at 12:15. If you get in it, you're forgiven." She pulled away and clapped along with the crowd.

As if someone were listening, she said loudly, "Goodnight, Major Benson."

"Good night, Senator. Thank you for the dance." Ron turned quickly and headed for the bar.

He checked his watch. It was 11:30. For some strange reason, he didn't feel like he was on the clock with her anymore. He didn't care if she paid him or not. Was he beginning to have feelings for her?

Ron drank slowly as he passed the time away. He watched couples and strangers as they mixed and mingled. As he looked across the bar, he saw Dorothy Henderson in a small group of people, staring at him. She held up her glass as a sort of long-range toast. Ron grabbed his wine glass, and did likewise. He took a drink and it hit him.

She knew. Dorothy was a sharp lady. She knew that he had been with Denise. A chill ran up his back and he began to think about how many people knew who he was and what he did in his off hours. Clyde knew, Dorothy knew, Denise and her security team knew. He wondered to himself, *how many other people knew?* Even Tony Parkman had heard 'rumors' about what he was doing. His private life may as well been on the cover of People Magazine. It was only a matter of time until the wrong person found out. It was only a matter of time until he was discovered and the wall of secrecy would crumble.

He checked his watch again and ordered a shot of tequila. The charade couldn't continue forever. He hadn't considered how the situation would end. Could he just quit and never do it anymore? Would he end up in Leavenworth? Was there a middle ground where he could just gracefully discontinue his work as he was now viewing the nightly visits?

Ron picked up the shot and looked at it. He was asking too many questions. He had an appointment in ten minutes. He licked the back of his hand and poured some salt on it, then picked up the shot and looked at it. Since his college day's he was always leery of finding a worm in his tequila. He licked the salt, poured the shot down quickly and bit into his lemon. It didn't help. The shot burned as it hit his stomach. The fire managed to burn away all the negatives thoughts that were racing through his mind. It actually helped to clear his vision. He tried to think about all the questions he had previously been asking himself. Screw it. He knew exactly what he was doing.

He was going to be a soldier. What would a soldier do? A soldier would suck it up! No more self-pity. No more fear. It's out there. Some people know. He made the choice. He wanted a change in his life and he was changed. He was alive.

Ron smiled at his own bravado. Underneath it all was a tiny voice replaying three little words. *You are screwed.* Ron shook his head and waved goodbye to the bartender. He gathered himself and headed to get his overcoat.

While he was waiting in line, he noticed himself in a hallway mirror. He couldn't help but see the uniform still fit him well. The couple pounds he had lost helped. He still looked like an officer on the outside. Inside, he knew the officer he was, was long gone. His change was more than physical. Deep down he knew it was the last time he would ever wear his blues. And it didn't bother him one bit.

The waiter was standing behind a pine tree outside so as not to be seen. He pulled out a large camera and snapped several pictures. He zoomed in and got the license number on the Escalade that Ron Benson climbed into. *Not a cab. Oh, yeah the General was gonna love this!*

CHAPTER 22

24 DECEMBER
145 PIEDMONT TRAIL
STAFFORD, VIRGINIA

"I think she's happy you came over tonight."

"I have some free time and in spite of what you think, I do love you both."

Sandy blushed and turned away at Ron's comment. "I know that."

"Just because we drifted apart, doesn't mean," Ron searched for the right words that didn't come. He never really could talk to Sandy.

She helped him. "I understand, Ron. It doesn't mean you can't still want to be with both of us." He smiled weakly. "I just couldn't live that life anymore. You would be on the road or off to war. The little officer's wives games. A daughter looking for a father that was rarely around. Even when you were here, you were never around!"

Ron nodded in silent agreement. He'd heard it before, but didn't feel like listening to it again. His holiday spirit seemed to be disappearing. He decided to change the subject. "I hope the money I sent for her college helps that some."

"Oh she was overjoyed with that little gift. Can't say David was happy about it."

Screw David thought Ron. He should do the right thing and marry Sandy so they could both stop living off Ron's monthly check. "So David knows I gave her the money." That didn't sit well with Ron, but he should have known she would tell him. "How do you feel about it?"

Sandy thought before she answered. "I don't know, Ron. How did you come up that much money? I know exactly how much you make. You couldn't possibly save that much."

Just then, David entered the room. Ron didn't like David very much. Ron always pictured him with a toothpick in his mouth and a sleeveless, white, wife-beater T-shirt. David wasn't that bad, so Ron gave Sandy the benefit of a doubt about David. Ron didn't acknowledge David and focused on Sandy when she said, "If I didn't know better, I'd swear you were doing something illegal."

David injected his perspective into the conversation. "You two talking about the money for Ally's college?" David knew about the money Ron had sent for Ally. "Where did you get that from? Did you deploy or something?" Ron didn't know how much of David's questions were envy and how much was jealousy. David's "career" as a truck driver was over due to chronic back pain. David often wondered how Ron was still healthy enough to stay in the military. Must be working in the Pentagon paid more or something. David didn't know shit about the military.

"No, I didn't deploy. Hell they won't let me." He didn't have a good answer. "I've just been scraping my dimes and quarters. She's still my daughter and she deserves whatever I can do for her."

David's true feelings came out. "Maybe you can scrap a few more dimes and quarters for Sandy." Ron started to get angry. Sandy could tell Ron was holding back. She'd only seen him angry three, maybe four times when they were married. His temper usually appeared when people questioned his decisions or the military. Whenever he got mad, it was justified. She wished David had that quality.

"The fact that he's given that money to Ally is enough. We're doing fine, Ron."

Ron looked at David and bit his tongue. "I think I'll go say my goodbyes to Ally."

"Leaving so soon?" asked David.

"Yeah, I'm gonna get to bed before Santa comes." Ron's retort was all he could come up with. The truth was he didn't want to answer anymore questions about where the money came from. Not from Sandy and certainly not David's.

Ron bent to kiss Sandy goodbye. He mustered up the strength to shake David's hand. "Merry Christmas."

David said, "See you next Christmas."

Ron thought about it. "Probably not." The response took David and Sandy by surprise. Before Sandy could ask what he meant, Ron left to say goodbye to Ally.

Sandy turned and smacked David lightly on the arm. "What?" he replied.

"What are you saying things like that for?"

"I don't like him popping in here and bringing himself into your lives like this."

"He's her father, David."

"Maybe so, but this isn't right. Something is going on with him, Sandy. Where'd he get that money?" He didn't add that he intended to find out what it was. If that meant talking to somebody at the Pentagon, by God he was gonna do it. If they were paying him more money, then Sandy deserved to get some too.

David Weston was too slow to understand that if the goose was killed no one got any eggs.

CHAPTER 23

2 JANUARY

2E800

PUBLIC AFFAIRS OFFICE

PENTAGON

Colonel Henderson was nervous. He didn't like the two agents that had just been in his office. They weren't exactly clear on why they wanted to see Ron Benson, but anytime the Criminal Investigation Division came around, the had their own reasons. None of them good.

As far as he knew, Ron had been doing a little better work recently. What was it he could be doing? *This is just what I don't need right now.* Just when things at home were getting better, something like this comes up at work. *I need to talk to Ron about this ASAP.* He called Ron's number and got no answer. He decided to have the Major come to his office. Rather than use his secretary, Henderson grabbed a folder off his desk, wrote a quick note and headed down the hall.

After the New Year, Ron showed up for work rested, yet cautious about his activities. He noticed most of his co-workers seemed to be different. Stand-offish was an appropriate way to describe their behavior. Even Tony Parkman seemed quieter in the cubicle. Was it the ball and him dancing with the Senator? That had to be it. Jan Stockman didn't even say 'Hi' in the hallway.

Ron looked at a folder filled with slides on his desk that had a note on them from Colonel Henderson. The note said, 'Come see me'. He looked through the slides and saw no changes were necessary.

Ron gathered up the slides and started toward Henderson's office. It was at that point when Ron, either by luck or fate, ran into Chris Watson, the civilian contractor. "You got a second?" Ron nodded. 'Yoda' pulled him aside and looked around.

"You headed in to see Henderson, right?"

Ron held up the slides and said, "Yeah."

"I don't think that's what he wants to see you about?"

Ron's eyes narrowed. "What do you think he wants to see me about?"

Chris looked around nervously. "I think some CID agents were in to see him first thing this morning." He looked toward the other end of the hall, then back at Ron. "You have something going on that I should know about?"

Ron was absolutely sure there was nothing going on the he needed to tell Chris about. "Nope." He followed Yoda's gaze. "Is there something you think I should know about?"

'Yoda' Watson blew off the evasive response and got to the point. "Those agents came in and were asking questions about you.

Ron got uncomfortable. He had expected someone, sometime would be coming to ask him a whole bunch of questions. It still made Ron sick to his stomach to actually know the Criminal Investigation Division was in the area and asking questions about him. He did his best to hide his fear. "You sure, Chris?"

Yoda looked over the top of his bi-focals and couldn't hide his Texas accent. "I'd bet your paycheck on it, Bubba. They want to know when you would be in and if they can talk to you?"

"What'd Henderson say?"

"He was hesitant to start. I don't think they gave him any information at first. But they closed the door and talked for about fifteen minutes." He looked down the hall again. "I think they changed his mind by the time they came out. Soon after that, he hand carried those slides down to your desk."

Ron nodded. "I guess it's time to go find out what this all about. Thanks for the heads up."

Yoda was confused by Ron's apparent carefree attitude. "That's it? You don't seem too worried?"

Inside Ron was an anxiety attack waiting to come out. He was surprised that Yoda hadn't noticed. He was doing all he could to keep from throwing up. "I need to find out more about what Henderson wants." He turned and started toward Henderson's office. Over his shoulder, he said, "Then maybe I'll get worried." As he headed down the hallway, he acknowledged the lie. He was scared as hell.

When he got to Henderson's office, the Colonel got up from behind his desk and motioned for Ron to take a seat. Ron hoped that Yoda was wrong and the get together was about the slides. He knew better.

Henderson closed the door and got down to business. "I had a visit this morning from two CID agents. They asked a whole bunch of questions about you. Wanted to know your off duty habits. Where you hung out? What you did? I didn't tell them much."

Ron focused on the word 'much'. He wondered what exactly that meant, yet remained silent.

"They asked me questions about whether you lived a flamboyant lifestyle or spent a lot of money? Quite frankly, I don't have a clue what they are talking about. I told them as far as I know, you spent a lot of time working late hours here at the office working projects for me. Is there something you want to tell me?"

To Ron it appeared the agents had not told Henderson what they were actually investigating him for. He assumed Henderson still had no clue about Ron's off duty life. "I don't think so, boss." It was a true statement.

Ron tried to change the subject. "Is there something wrong with the slides?"

"Slides?" Ron held them up. "Oh, no. Not at all. That was an excuse to get you in here early. I didn't want the rest of the office to know what CID was doing here."

Ron let it slide that the office already knew. That's why everyone kept their distance from him. "Thank you for that, sir."

"You sure you don't know what they want to talk to you about?"

Ron shook his head. "No, sir."

Henderson nodded. "Expect to talk to them later this afternoon. They'll be back."

Ron left the office and could feel the eyes of his co-workers on his back as he headed back to his cubicle. He sat down and exhaled loudly. As his stomach churned, he kept reassuring himself that everything was going to be all right. The agents didn't have anything on him. *Maybe they wanted to talk about something else.*

Yet Ron was a realist. He knew what was going on. It was just a matter of time. Tony was out, so he sat in his cubicle alone, waiting for the call that would probably end his career.

CHAPTER 24

2 JANUARY
2E800
PUBLIC AFFAIRS OFFICE
PENTAGON

Ron got the call that told him he was to meet two CID agents. They would meet Ron in his cubicle at 1600 hours. Ron thought about making some excuse not to make it there, but could find no good reason to shirk the meeting. While waiting to meet the agents, he made a decision. He may be headed out of the military, but he wouldn't go without making it difficult, especially for the agents.

"We have a couple questions to ask you, Major Benson." The CID agent showed Ron his badge. Agent Kevin Crocco was a Warrant Officer who started his career as a Military Policeman. At 37, he was the older investigator.

His junior partner was an attractive younger woman of 25, with short brown hair and no sense of humor. Angelica Ramirez must have just finished her training as an agent. She let Agent Crocco do the questioning and never once smiled.

"I have no problem with a couple questions, Agent Crocco."

"First of all, do you know an individual named Richard Head?" Ron shook his head. "How about Lance Longman?" Ron was detecting a pattern. Again, he shook his head no.

It was agent Ramirez that asked, "Does the name Clyde Simpkins mean anything to you?"

Ron looked at Agent Ramirez and then back to Crocco. He saw no value in lying. "I know Mr. Simpkins."

"All right. We received a tip from a Mister David Weston that you recently sent a check for $12,000 to your daughter. Is that true?"

Ron nodded, *Fucking David!* That was something he hadn't counted on. "Yes, I did. It's for college next year." *If I go to Leavenworth, does Ally get the money?*

"We're wondering how you managed to save that much money?"

"I eat a lot of peanut butter."

Agent Ramirez stepped forward and handed Crocco a folder. "You're a smart guy, Major Benson. Maybe you can help explain these?"

The folder contained some pictures that Agent Crocco pulled out and handed to Ron to take a look at. "Do you recognize the woman in the photo?"

The picture was of Dorothy Henderson sitting at the bar. The next picture showed the two of them together. Again, Ron nodded. "Mrs. Dorothy Henderson."

"She's your boss's wife, isn't she?" Crocco was starting to get too smug for Ron.

Ron answered flatly. "Yes."

"Would you like to tell us what you were doing with her in that bar?"

"Nope."

Agent Crocco was visibly pissed. "How about some of these other pictures?"

Ron glanced at some of the pictures. He recognized the women, almost all of locations of the pictures in bars, restaurants and a couple near the women's residences. Ron remembered when most of them were taken. For the most part, the shots were taken in the last month, when he was a more comfortable with his second job and more public with his companions. His stomach churned in anticipation of seeing his most famous client. He was surprised when no pictures of Senator Denise Mitchell-Samuelson appeared. *Maybe they didn't know about her.* He laid the pictures down and said, "Do I need a lawyer?"

Agent Crocco was still coy. "You tell us, Major."

"I'm not going to tell you anything, Agent Crocco. I believe I will be contacting the Judge Advocate General's office before I say anything else."

"That's fine with us, Major. We'll just be going back to see Colonel Henderson with our evidence and see what he thinks. We've got a large some of money from somewhere. We've got you with multiple women, on multiple evenings. That's just the beginning.

"For some reason, Colonel Henderson thinks you're a good officer." Crocco looked Ron straight in the eye, with an omnipotent smile said, "That picture of you and his wife ought to make his opinion of you change." Ron bit tongue and tried to show no emotion. Ron was certain Crocco could see his anger.

Agent Ramirez continued, "I imagine he will encourage us to see General McCoy to start criminal proceedings. You can expect to be in his office first thing tomorrow."

Ron knew better than to be antagonistic to the agents. They were doing their job. "We'll cross that bridge when we get there, Agent Ramirez. So if you two can find your way back to Henderson's office on your own, please do. I need to call the JAG."

Immediately after the agents left, Ron made a phone call. It wasn't to the JAG office, it was to Clyde Simpkins. He could only leave a page number to have Clyde call him back.

"This must be something important to have me call you at work?"

"I just got through with a visit by two CID agents. They're probably listening."

"Did they give you enough information to know what they were after you for? How bad is it?"

"Let's see if I get these right; 'Richard Head', 'Lance Longman'? Do those names ring a bell?"

"Crap."

"They have pictures, too."

"Do they have pictures of all your," he searched for better words in case their conversation was really being taped. "friends?"

"I think they only showed me a partial collection."

"What are you going to do?"

"I'm going to wait until they charge me with something then get a lawyer. They told me I would be getting called into General McCoy's office tomorrow. Pretty sure I'm getting a new set of orders. Probably a long tour at the Federal Crowbar Hotel in Kansas."

Clyde was silent for a moment. Long enough to make Ron uncomfortable. Finally he said, "I'm going to see what I can do to help you. I won't make you any promises, but I'll try."

"I don't know what you can do, Clyde. I only called to give you a heads up."

"I'm pretty certain they aren't after me, brother. I think it's you they want. An officer in the military can't be doing the things you're doing without pissing off the system. They're gonna try and make an example out of you."

"I believe that. What are you gonna do?"

Clyde was evasive. He didn't want to lie to his friend, but he didn't want anyone else to hear his plans. "I can't be a witness, Ron. For or against, you know?"

Ron was quiet momentarily. "Okay. That seems fair enough. No help, but no hindrance either."

"Look, I'm gonna see what I can do. You will hear from me again. Someday." The phone went dead.

Ron Benson had never felt so alone.

2 JANUARY
2310 WASHINTON HEIGHTS
ALEXANDRIA, VIRGINIA

Clyde thought hard about what Ron had told him. He had expected something like this could happen. That was the price he paid for taking a partner that was in the military. Clyde was hoping he would get a few more months before he had to leave.

The first thing he did was get on his cell phone and called number 29 in his address book. As he waited for an answer, he grabbed his largest suitcase and threw it on the bed.

A female voice answered the phone. "There is a situation developing that I think you need to know about." It only took him two minutes to explain. Clyde packed quickly as he talked. As soon as he hung up the phone, the reality hit him. It had been a good run in DC. It was time to move on. He had done the only thing he could do for Ron. His situation was in someone else's hands.

CHAPTER 25

2 JANUARY

2E800

PUBLIC AFFAIRS OFFICE

PENTAGON

R on felt sick. He didn't want to answer the phone. He let Tony Parkman answer all the calls. Ron's mind raced as he waited for the unknown.

Tony Parkman left Ron alone. He knew the CID had come by and interviewed Ron. They kicked him out of his own office to talk to Ron. He also knew his friend was in deep shit. He didn't know what the interview was about. Tony was absolutely positive Ron couldn't have been doing anything to get in trouble.

Then something popped in his mind. The rumors about a prostitute in the Pentagon. *Maybe Ron knew something about that? Could it be Ron was the guy?* He looked over his shoulder and saw his nervous friend. *No way! Maybe he knows who it is?* Tony was certain no one would pay Ron Benson for sex.

Finally, the bell tolled. At four fifty in the afternoon, he received an email from General McCoy's secretary. He was report to the General's office at nine o'clock sharp the next morning.

3 JANUARY
DIRECTOR OF PUBLIC
AFFAIRS OFFICE
THE PENTAGON
WASHINGTON DC

Ron Benson hadn't slept much. Negative thoughts filled his mind. Visions of life in a small prison cell in Kansas kept flashing through his mind. He woke up after three hours of restless sleep and slowly got ready for work. He dressed in his Class A Green uniform and headed into work.

Ron showed up ten minutes early at McCoy's office. Agents Crocco and Ramirez were there in the office waiting. Ron just nodded at the agents and took a seat. The secretary went into the General's office and returned promptly. The General wanted to see all three in his office at the same time.

Ron didn't bother to report. The General directed all three to take seats around a coffee table. The General seemed to be unconcerned with the threesome as they sat down. Agent Crocco tried to hand him the folder of photos, but the General pointed to the coffee table. "I saw them last night." He addressed Ron. "Agent Crocco seems to have some pictures of a disturbing nature Major Benson. He has explained to me that you were not inclined to discuss the photos with him."

"That's correct, sir."

"Why is that?" asked the General.

Ron was hesitant. He could see Crocco glaring at him. Ramirez was following the senior agents' lead. "I'm not sure what he's doing with them, General. Neither he, nor agent Ramirez has indicated to me what this is about."

"Did you go to the JAG, Major Benson?" asked Crocco.

"I'll ask the questions in here, Agent Crocco." The General was leaving no doubt about who was in charge. "You were saying, Major Benson?"

"I was saying, the agents have made some assumptions about what I do in my spare time and have alluded to ill gotten funds, but have not charged me with anything. I have not gone to the JAG office because I'm not certain what they want from me?"

Crocco couldn't take it anymore. He couldn't stand Ron Benson! He knew he had Benson in his sites and this whole charade with the General was a formality. Crocco didn't like officers, but ones that broke the UCMJ, he hated. This was more than he could take. Crocco jumped up and yelled, "You know damn well what we want from you! You need to explain what's going on in your life Major!" He took a step towards Ron. "You're a damn disgrace to anyone that has ever put on a uniform!"

Ron stood up and met Crocco's advance. "What do you know about wearing a uniform? You're just an undercover cop that gets your kicks from sellin' out the guys that really wear the uniform! You couldn't handle the real Army, so now you waste your time go after real soldiers!" Ramirez stood up and got between the two men.

The General had enough. He snapped, "AT EASE! Both of you! He stood up and directed them to take their seats.

Crocco couldn't let go. He turned to the General and said, "This guy is in violation of at least five codes of the Uniform Code of Military Justice. We've got adultery, probably tax evasion based on the money and we're thinking that might lead us to prostition." Ron was silent. "Who knows how many other things."

Suddenly across the room, the door to the General's office opened and his secretary appeared. She hurriedly came across the room and pulled the General aside.

Ron sat back in his chair and watched as the secretary whispered in his ear. Apparently someone was there to see the General. He heard a knock on the door and a familiar voice say, "I hope I'm not interrupting anything, Patrick!"

Senator Denise Mitchell-Samuelson entered the General's office as if they were old friends. "Hello, General McCoy! It's been so long since I've come over to visit."

General McCoy walked over quickly and extended his hand. "Good morning, Senator. How nice to see you again."

"I hope you don't mind if I stopped by." She looked around the room and tried not to stare at Ron. "I was hoping we could talk."

"Certainly, Senator. I think these folks can wait out side." The General looked at Ron and the Agents. They were already moving toward the door.

Ron stole a glance at Denise on the way out. She had moved to the fifth floor window and was looking at the river. The General closed the door behind them.

Ron sat down heavily into a chair and grabbed a magazine.

"What kind of bullshit is this, Benson?"

Ron looked up at the agent and said calmly, "Sit your ass down and wait until the General wants to talk to us." Crocco was visibly pissed, but did exactly what the superior officer said. He may have been a criminal, but until he was proven guilty, Ron still outranked the agent.

"All right, Patrick. I'll come straight to the point. What do you know about this situation with Major Benson?"

McCoy didn't become a general because he was a fool. The fact that she was in his office, the same time this "Major Gigolo" nonsense was starting to surface could not be a coincidence. He'd seen the picture of Benson with her. Saw the limo register to her with Benson getting in. *She's here to save him.* As an officer, Benson had been above average. Benson was on the money with his military assessment of PA and the War. But it was the attitude that the General couldn't stand. The air of cockiness. The audacity to think he had it all figured out. McCoy had the War figured out way before Benson came into his office whining. *Not so cocky now, eh Major.*

Now the Senator shows up. *I don't have time for any of her shit.* The General wanted to get straight to the point. The pleasantry's aside, Senator Mitchell-Samuelson was usually a full-fledged bitch whenever he testified before her committee.

McCoy walked up beside the Senator and followed her gaze outside. The January weather was taking a break and the sun was trying to shine. "I know everything I need to know, Denise."

The Senator glanced sideways at the General. By calling her by her first name, he was letting her know that he understood the situation better than she expected. She was gonna owe him big time.

"I can't have this get out to the press. If you're the only one that knows, there is a slim chance my political career isn't over."

McCoy, still looking out the window, smiled broadly. "You know with another star, I could have a better view than this."

Denise didn't hesitate. "That shouldn't be a problem, Patrick."

"I think some Joint time out in Hawaii might be just what I need to set up my retirement."

She turned on him. "Don't push it! Fuck with me, Patrick and you'll have your retirement sooner than another star." She glared at him, then turned sharply and walked in front of his desk. "I seem to have recently come up with a little dirt on your early years as a lieutenant. Seems there was an incident involving a young female private. Was she pregnant, Patrick?"

McCoy was hot. But knew enough to hold his Irish temper. He moved behind his desk. "Touchez, Senator. What do you want?"

"I'm guessing there are pictures. Of me. Of others."

"Dozens."

"I want them. I want them all. Some of those women are friends of mine. We can't have rumors like this going around town."

The General picked up the folder. "Perhaps more friends than you expect." He let the comment sink in. Then added, "Some of the women you may not know, Denise. But he knows them." He handed her the folder. She opened and looked through the pictures. Although disturbed by the photos, she didn't want McCoy see it.

"That's an impressive portfolio your boy picked up, Denise. Bet you didn't know he was sleeping with half of DC, did you?" McCoy watched her eyes as she looked through the pictures. Every once in awhile, he thought he could tell that she was impressed by Benson's clients. *Perhaps it was jealousy?*

"This is valuable information in more ways than one, General."

"The military is here to serve the government, Senator." He got back to his point. "What else do you want?"

"We can't very well do anything to him, or this," she held up the pictures, "becomes public. It's not just me here, it's some of the most influential women in Washington. Not to mention the black eye this whole gigolo thing brings to the military."

"What do you think he'll want?"

"I'm guessing he just wants it to go away. Maybe to stay in long enough to retire. Is that possible?"

"He only has a year or two left. He's not going to get promoted. Considering his option is Leavenworth, he'll take the deal."

"I think he wants to be with soldiers."

"No. No, I may have had problems early in my career, but that sorry piece of crap will not be with troops at the end of his. I'm gonna send him so far north, it'll take a microwave to defrost his balls."

"Sounds interesting, Patrick. That seems a very appropriate outcome."

She nodded and stuck out her hand. "What about the CID agents?"

"I'll handle them, Denise."

She turned and said, "Patrick, let's never talk of this again."

McCoy nodded as he showed her to the door. He started to kiss her hand and she quickly pulled it away and opened the door. "Thank you, General. Until we meet again."

McCoy stepped into the waiting area. "Thank you, Senator." She walked by and stopped about ten feet from Ron.

"Agent Crocco, Agent Ramirez, could you come into my office please." Crocco gave Ron an angry stare and followed Ramirez into McCoy's office.

She stood there looking at Ron as the door closed. Ron noticed the folder of photos was in her hands. "I hope that's all of them."

She smiled. "Oh, it will be. Whatever they have, the General will get for me. I don't think he would appreciate my nasty side if some were to turn up on the internet."

Ron smiled and looked down at his feet like a little kid. Ron's eyes couldn't cover his fear. His voiced cracked a bit when he asked her. "Am I going to jail?"

"No, not today. We have an arrangement. We can't very well have you making a scene with all the, let's call it, information you have accumulated." She held up the pictures. "And you've accumulate a lot, haven't you?"

She shook her head. "Some of these women are my friends! I mean, how could you?" She was trying to laugh when she said it. "I understand how you could, but I . . ." She couldn't finish. It was at that moment that Ron thought maybe she had feelings for him. She changed her focus.

"I believe you are leaving DC, Major Benson."

Ron's eyebrow went up. "Is he sending me to Iraq?"

"I don't think so. Something of frost on your nuts I think he said?"

Ron smiled. "At least it's not Leavenworth."

"One more thing, Major Benson." She walked over, kissed him quickly on the cheek and dropped the hammer. "I never want to see you again. Do you understand?"

So much for thinking she had feelings for him. "Yes. Yes, I do, but . . ."

She turned quickly less he see her tears starting to form. "I've got a campaign to get ready for and I don't need to hav any distractions around." She stopped at the door, but did not turn around. "Goodbye, Major Benson." She quickly opened the door and left without letting Ron see her tears.

Some of it was anger at her own foolish behavior; some of it was rage from her jealousy. Yet, some tears were for her unfulfilled desires to be with him. *Damn it! How could I have ever let myself care for this guy? It wasn't love. It was never love..*

Ron swallowed hard fighting his own emotions. She was gone. He didn't have time to thank her or even say 'Goodbye'. Not even one last kiss.

He didn't even have time to think about her quick departure as the General's door opened behind him. He took a deep breath and turned around. Ramirez came out frowning, closely followed by Crocco. His face was flushed red with anger.

"Agent Crocco make certain I have all the information on my desk tomorrow morning," said McCoy. He motioned to Benson to come in to his office. "Goodbye agents and thank you."

As Ron walked by Crocco the agent hissed, "You don't deserve to be in my Army, you piece of scum."

"You're not in the Army, agent Crocco. You're a cop." He turned and said over his shoulder, "There's a Dunkin Donut on the corner, why don't you wait for me there."

Crocco turned to follow Benson, but Ramirez grabbed him and the General was waiting at the door with a disapproving scowl.

"Sit down, Major Benson." The General took the large chair that faced the coffee table. Ron had no expectations, only the hope that he would leave the room with some semblance of dignity.

"This is quite a situation you've put us in. You've a got a senior Senator that you've put in a compromising position. I'm wondering if you blackmailed her. You've embarrassed the Army and you've broken, God only knows how many laws."

"I'm sorry, sir, I . . ."

The General put his hand up and bit his lip. He was doing a terrific job of holding his temper. "Just listen to me." He breathed in heavily and started again. "Now I've done everything I can to keep those two agents from going over my head. And I'm not so sure they won't. That's why you need to be out of here."

"How long do I have?"

"Hell, son, you were gone last night and didn't say goodbye! Before you can get back to your desk, I will have orders cut for you."

"Where to, Sir? Any chance I can be with troops?"

"Troops? Why the hell would I give you the honor of being with troops? So you can go fight? Nobody wants to fight with scum like you, son." Ron started to feel his temper rise. He bit his tongue and squirmed in his chair.

"If I had my way, your orders would say Fort Leavenworth, Kansas for a long tour. Like maybe 25 years. After you blackmailed the Senator and slept with your bosses wife, I have absolutely no respect for you."

Ron jumped up. He couldn't take it. "You don't respect me? You don't respect me! How the hell do you respect yourself?"

"Just a minute, Major!" The General was glaring at Ron. He started to get out of his chair.

Ron cut him off and motioned for him to stay put. "No, General! It's my turn! I took your bullying, but I won't take your shit!" The General started to get up again. "JUST SIT DOWN AND LISTEN!"

Ron walked over towards him. "I did NOT screw Dorothy Henderson! Never! I saw her in the bar where the pictures were taken, had a drink and went home. From what I hear, their marriage is much better now."

"As far as the Senator is concerned, I don't know what she was doing here, because I didn't call her. I don't even have her number. She calls me. So how could I blackmail her, Sir?" He stepped away and offered his own assumption. "It also seems she knows you, Sir. Does she have dirt on you?"

"You scum!" hissed McCoy.

"Yeah, I'm scum all right. I've fucked around some, but I haven't fucked anybody over. How long have you been in the Pentagon, General? If you've been here a day, you've been here too long. The minute you take a seat here, you stop fighting the enemy and start fighting congress. Remember them? They're the people we work for. You've forgotten who and how to fight, so you have your staffs kill electronic forest in the hopes they can get you recognized, promoted and sent back into the big game." Ron shook his head and looked at the General. "Sure, I'm scum. But I have never forgotten where I came from. I might not be a good one, but I'm still a soldier."

McCoy was pissed, but the words hit home. "I've half a mind to call her back and say the deal is off!"

"You won't! You know you won't! What ever she has on you or offered you, you'll take because you've lost your compass as much as me!" Ron's eyes were tearing up. He walked over to the window and looked out hoping the time would help him get his composure.

It didn't. "I didn't ask to come to this fucking place. The Department of the Army sentenced me here. I wanted to stay with troops. I tried to go to Iraq, but they said I was not needed to fight." He turned and looked at the General. "I was needed to make slides and briefings that look good. I was needed to write papers. Do you understand how degrading that is for someone that wants to fight? Do you?"

McCoy had moved next to him. They both looked out the window. The sun shining over Washington on a clear day early in the morning had a calming effect on both men. Ron turned away and wiped at his eyes, hoping the general didn't see him. He wasn't even looking.

"Yes, I do." The General took in a deep breath. "I remember those days. You probably don't believe this, but I tried to get there too." Ron chuckled softly. "I did. They said I was too old. So I'm up here in this penthouse office." He turned back and walked toward his desk. "Might as well be a pasture." He sat down heavily.

Ron turned and walked directly in front of the General's desk and stood straight. He knew he was out of line and said what he only hoped would be seen as a conciliatory gesture. "Sir, I would like to apologize for my behavior. Not just for the last few months, but for the outburst just now. I'm truly sorry."

"Relax, son. We're both sons of bitches."

Ron came to the position of parade rest. McCoy felt awkward, almost uncomfortable. He could tell Ron Benson did want to be a soldier. McCoy couldn't blame him. He would have traded the new star Denise had promised him for a chance to lead troops again. That's all Benson wanted. If he had been with troops, he wouldn't have gotten into trouble he was in. *I can't hammer this guy.* "What do we do with you?"

"I'd still like to serve out my twenty, sir."

General McCoy tapped a pencil nervously on his desk. "And you won't fuck up, literally, anymore, Major Benson?"

Ron smiled. "No, sir. I'm done with that if you give me the opportunity to retire."

MCoy came up with an idea that he thought might work for Ron. McCoy was no fan of Benson, and he felt the Major probably should be headed to Leavenworth. If he could go far away, the whole event would go away with him. "I've got a great friend of mine. He lives in Montana. He's the Adjutant General of the National Guard up there. We go hunting sometimes. I think he needs an active duty Major to help his National Guard armor battalion get trained up. Should take a year or so."

"Montana, Sir?" Ron nodded. *It ain't Mongolia.* "That would be great, Sir."

"You're damn right it would be! Considering your option is jail!" Ron was silent. "You think you can lay low and stay out Crocco's way for 48 hours until the paperwork catches up and your ass is gone?"

"Yes, Sir."

"Sleep in your own bed, too."

Ron suppressed his smile. "Yes, sir."

McCoy sat back heavily in his chair. "Now that that is all over, I have a question for you."

Ron relaxed. "Yes, sir."

"I saw the photo's that Crocco had. I have to admit, I was quite impressed. I don't know how you got into that line of work or why you did it. But I do respect you as the soldier you were, son. I have one question for you." Ron smiled cautiously. He wasn't sure if he was getting set up, so he swore to admit nothing. Ron was afraid McCoy would ask how Senator Mitchell-Samuelson was in bed. One of Clyde's best pieces of advice was to never talk about customers' quality, experience or lack of it. That was like doctor patient privilege. Not to be shared. He needn't have worried.

"If you can tell me, how much did you make?"

Ron was stunned. "How much money, Sir?"

"Yeah, how much money did you make?" Ron wasn't hesitant to answer. "I saw some of the women and quite frankly, there are some beautiful women there. I was just curious about how much they paid you. I won't tell anyone."

Ron smiled broadly. "I made over twenty-five thousand dollars in just about three months, sir."

"Damn, son!"

"What can I say, General. I enjoyed my work."

McCoy shook his head. "Get out of my office. I never want to hear about this again."

Ron stood straight and saluted crisply. The General watched as the Major held his salute. He extended Ron the courtesy of returning it. "Hoo-ah, Major. Get the hell out of here."

Ron completed the salute and quickly exited the room. McCoy looked out the window and wondered how the Army let soldiers like Benson get so messed up. He grunted as the answer dawned on him. They sent them to the Pentagon.

EPILOGUE

27 MARCH
MONTANA ARMY NATIONAL GUARD CENTER
BOZEMAN, MONTANA

"Hey, Major Benson!" came the yell from across the room. The young Specialist held up the phone and said, "There some guy here on the phone from California. Says his name is Dickman. He wants to talk to you."

Ron was puzzled for an instant. He didn't know anyone in California. Then it hit him. A smile appeared on his face. "I'll be right there." He excused himself from the soldiers he was talking to.

He grabbed the phone. "Major Benson."

"So, you're still a Major, eh?"

The smile broadened. "Somehow. Yeah."

"What's the weather like up there?"

"Ten degrees and snowing."

"I'm looking out over a swimming pool. There's about fifteen beautiful ladies here. I've got a tequila sunrise in my hand and I'm about to order some sushi."

"You've gone native, Clyde. Or whatever your name is now?"

"Clyde still works."

"I'm amazed that you found me."

"It wasn't too hard. The internet is a marvelous thing. You're actually the third Major Ronald Benson I've called. I'm so glad you're not in Leavenworth."

"That makes two of us. What can I do for you?"

"I need some help!"

Ron was silent for a moment. "What kind of help?"

"I'm overworked, brother! Is there anyway you can get down here?"

Ron hesitated. "I don't know Clyde. Last time we hooked up, I almost went to jail."

"But it was a hell of a ride wasn't it?"

Ron nodded as if Clyde could see him. "Yeah. Yeah it was."

"Look, I'm in Beverly Hills. Business is picking up. I'm serious about the weather. It's 85 degrees here." Ron looked out the closest window and saw the snow blowing. "You're almost retired aren't you?"

"I've got less than a year left."

"That's perfect! I can have things on this end in full swing by the time you get down here."

Ron was silent again. A Sergeant came over with some papers for him to sign. He looked them over and smiled broadly. "Clyde can you hold on and I'll get your number." He signed the paperwork and sent the Sergeant back to work.

Clyde gave him his number and said, "Now I'm not holding you to anything. I know you don't want to make me any promises. But I seriously have more than enough work."

"Let me think about it and I'll get back to you."

"You do that! See you soon, Major Gigolo!" Ron heard him laugh as he hung up.

Ron smiled, hung up the phone and looked around. Different soldiers were moving around the office, doing what they do. More papers had to be signed. The unit was getting ready to deploy to Afghanistan. Ron Benson would not be going with them. He looked at the snow blowing outside the window.

Ron walked back to his desk and picked up the phone. He pulled a phone number from his wallet and dialed. "Hi, this is Major Ronald Benson. I need to initiate my retirement paperwork."

The secretary on the other said she was sorry to hear that. Ron answered, "Sometimes you just reach a point in your life and you know it's time to go do something else."